PORTABLE

WORKSHOP™

Basic Wood Proje
with Portable Po

Kids' Furnishings

COWLES
Creative Publishing, Inc.

Minnetonka, Minnesota, USA

Credits

Group Executive Editor: Paul Currie
Project Director: Mark Johanson
Associate Creative Director: Tim Himsel
Managing Editor: Kristen Olson
Project Manager: Ron Bygness
Lead Project Designer: Jim Huntley
Editors: Mark Biscan, Steve Meyer
Editor & Technical Artist: Jon Simpson
Lead Art Director: Gina Seeling
Technical Production Editor: Greg Pluth
Project Designer: Steve Meyer

*Vice President of Photography
  & Production:* Jim Bindas
Copy Editor: Janice Cauley
Shop Supervisor: Phil Juntti
Lead Builder: John Nadeau
Builders: Troy Johnson, Rob Johnstone
Production Staff: Laura Hokkanen, Tom
  Hoops, Jeanette Moss, Gary Sandin,
  Mike Schauer, Mike Sipe, Brent Thomas,
  Kay Wethern

Creative Photo Coordinator:
  Cathleen Shannon
Studio Manager: Marcia Chambers
Lead Photographer: Rex Irmen
Contributing Photographers: Steve Smith,
  Rebecca Schmitt
Photography Assistant: Dan Cary
Production Manager: Stasia Dorn
Printed on American paper by:
  Inland Press
  00 99 98 97 / 5 4 3 2 1

COWLES
Creative Publishing, Inc.
Minnetonka, Minnesota, USA

President: Iain Macfarlane
Executive V.P.: William B. Jones
Group Director, Book Development:
Zoe Graul

Created by: The editors of Cowles
Creative Publishing, in cooperation
with Black & Decker. **BLACK&DECKER** is a
trademark of the Black & Decker
Corporation and is used under license.

Library of Congress
Cataloging-in-Publication Data

Kids' furnishings.
        p. cm.—(Portable Workshop)
  At head of title: Black & Decker
  ISBN   0-86573-682-0 (hardcover).

1. Children's furniture.
I. Cy DeCosse Incorporated.
II. Black & Decker Corporation (Towson, MD)
III. Series.
TT197.5.C5K54   1996
684.1' 04—dc20   96-17655

Contents

Introduction

You're browsing in the hottest new upscale childrens' furniture store and a cute little oak chair catches your eye: nothing fancy, just a good solid chair. Interested, you look more closely. You recognize the manufacturer, and you turn the label over. You see the price. "That costs more than the adult-size!" you exclaim. So you try another kids' store—a discount chain. The prices are much better, but everything you see is plastic: plastic bed-frames, plastic dressers, plastic toy chests. And all of it made in some faraway country with eight syllables in its name that you've never heard of before. How can you know what you're really buying? With *Kids' Furnishings*, an innovative new project book from the

Black & Decker® *Portable Workshop*™, you have a better alternative: build it yourself.

This book contains 20 fun, creative and functional kids' furnishings projects that you can build without a fancy workshop or a lot of experience working with wood—and without spending a lot of money. And by making these furnishings and accessories yourself, you will know exactly what goes into each project, and you will know that it was made with care and with pride.

So if you are trapped between paying exorbitant prices and settling for junk, take a look through the simple, creative wood projects in this book. Here, you will find useful projects that you can manage: from a simple oak bathroom stool, to a kid-size oak desk, to a fun and fanciful dinosaur rocker.

For each of the projects in *Kids' Furnishings*, you will find a complete cutting list, a lumber-shopping list, a detailed construction drawing, full-color photographs of major steps, and clear, easy-to-follow directions that guide you through every step of the project.

The Black & Decker® *Portable Workshop*™ series gives weekend do-it-yourselfers the power to build beautiful wood projects. Ask your local bookseller for information on other volumes in this innovative new series.

NOTICE TO READERS

This book provides useful instructions, but we cannot anticipate all of your working conditions or the characteristics of your materials and tools. For safety, you should use caution, care, and good judgment when following the procedures described in this book. Consider your own skill level and the instructions and safety precautions associated with the various tools and materials shown. Neither the publisher nor Black & Decker® can assume responsibility for any damage to property, injury to persons, or losses incurred as a result of misuse of the information provided.

Organizing Your Worksite

Portable power tools and hand tools offer a level of convenience that is a great advantage over stationary power tools. But using them safely and conveniently requires some basic housekeeping. Whether you are working in a garage, a basement or outdoors, it is important that you establish a flat, dry holding area where you can store tools. Set aside a piece of plywood on sawhorses, or dedicate an area of your workbench for tool storage, and be sure to return tools to that area once you are finished with them. It is also important that all waste, including lumber scraps and sawdust,

be disposed of in a timely fashion. Check with your local waste disposal department before throwing away any large scraps of building materials or any finishing-material containers.

> **Safety Tips**
> •*Always wear eye and hearing protection when operating power tools and performing any other dangerous activities.*
> •*Choose a well-ventilated work area when cutting or shaping wood and when using finishing products.*

Tools & Materials

At the start of each project, you will find a set of symbols that show which power tools are used to complete the project as it is shown (see below). You will also need a set of basic hand tools: a hammer, screwdrivers, tape measure, a level, a combination square, C-clamps, and pipe or bar clamps. You also will find a shopping list of all the construction materials you will need. Miscellaneous materials and hardware are listed with the cutting list that accompanies the construction drawing. When buying lumber, note that the "nominal" size of the lumber is usually larger than the "actual size." For example, a 2 × 4 is actually $1\frac{1}{2} \times 3\frac{1}{2}$".

Power Tools You Will Use

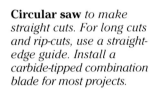

Circular saw *to make straight cuts. For long cuts and rip-cuts, use a straight-edge guide. Install a carbide-tipped combination blade for most projects.*

Drills: *use a cordless drill for drilling pilot holes and counterbores, and to drive screws; use an electric drill for sanding and grinding tasks.*

Jig saw *for making contoured cuts and internal cuts. Use a combination wood blade for most projects where you will cut pine, cedar or plywood.*

Power sander *to prepare wood for a finish and to smooth out sharp edges. Owning several power sanders (⅓-sheet, ¼-sheet, and belt) is helpful.*

Belt sander *for resurfacing rough wood. Can also be used as a stationary sander when mounted on its side on a flat worksurface.*

Router *to cut decorative edges and roundovers in wood. As you gain more experience, use routers for cutting grooves (like dadoes) to form joints.*

Guide to Building Materials Used in This Book

•Sheet goods:
AB PLYWOOD: A high-grade plywood usually made from pine or fir. The better (A) side is sanded and free from knots and other defects that require filling. Moderately expensive.
BIRCH PLYWOOD: A highly workable, readily available alternative to pine or fir plywood. Has a very smooth surface that is excellent for painting or staining, and generally has fewer voids in the edges that require filling. Moderately expensive.
LAUAN PLYWOOD: A lightweight wood panel sold in 4 × 8 sheets. Often used for floor underlayment. Has smooth surfaces, but little structural strength. Very inexpensive.
MELAMINE BOARD: Particleboard with a glossy, polymerized surface that is water-resistant and easy to clean. Inexpensive.
PEGBOARD: Perforated hardboard sold in ⅛ and ¼" thickness. Very inexpensive.

•Dimension lumber:
PINE: A basic softwood used for many interior projects. "Select" and "#2 or better" are suitable grades. Relatively inexpensive.
RED OAK: A common hardwood that stains well and is very durable. Relatively inexpensive.
CEDAR: Excellent outdoor lumber with rich warm color. Moderate.

Guide to Fasteners & Adhesives Used in This Book

•Fasteners & hardware:
WOOD SCREWS: Brass or steel; most projects use screws with a #6 or #8 shank. Can be driven with a power driver.
DECK SCREWS: Galvanized for weather resistance. Widely spaced threads for good gripping power in soft lumber.
NAILS & BRADS: Finish nails can be set below the wood surface: common (box) nails have wide, flat heads; brads or wire nails are very small, thin fasteners with small heads.
Misc.: Door pulls & knobs; butt hinges & utility hinges; metal L-braces; corner bracket; shelf pins; wood plugs (for filling screw counterbores); plastic glide feet; plastic drawer glides; other specialty hardware as indicated.

•Adhesives:
WOOD GLUE: Yellow glue is suitable for all projects in this book.
MOISTURE-RESISTANT WOOD GLUE: Any exterior wood glue, such as plastic resin glue.
EPOXY: A two-part glue that bonds powerfully and quickly.

Finishing Your Project

Before applying finishing materials like stain or paint, fill all nail holes and blemishes with wood putty or filler. Also, fill all voids in any exposed plywood edges with wood putty. Sand the dried putty smooth. Alternative: fill counterbored pilot holes with wood plugs if applying stain. Sand wood surfaces with medium-grit sandpaper (80- to 120-grit), then finish-sand with fine sandpaper (150- to 180-grit). Wipe the surfaces clean, then apply at least two coats of paint (enamel paint is most durable), or apply stain and at least two coats of topcoating product (water-based polyurethane is a good choice).

Covered Sandbox

*The cedar canopy provides shelter from rain and sun when raised,
and keeps out pests and debris when lowered.*

CONSTRUCTION MATERIALS

Quantity	Lumber
2	1 × 2" × 8' cedar
2	1 × 6" × 10' cedar
4	2 × 2" × 8' cedar
4	2 × 4" × 8' cedar
1	2 × 6" × 8' cedar
3	2 × 8" × 8' cedar
1	⅜" × 4' × 8' cedar plywood

Sandboxes are a backyard favorite for youngsters everywhere, but they do have a few natural enemies: rain, fallen leaves, backyard debris, and even neighborhood cats can make any sandbox uninviting. That's why we designed this roomy cedar sandbox with a sturdy cover that protects against these enemies when it is lowered over the top of the box. We took that good idea one step further by creating the cover so it can be supported by removable corner posts to do double duty as a sun and rain shelter when the sandbox is being used. When the cover is lowered, the posts can be stored neatly inside the sandbox.

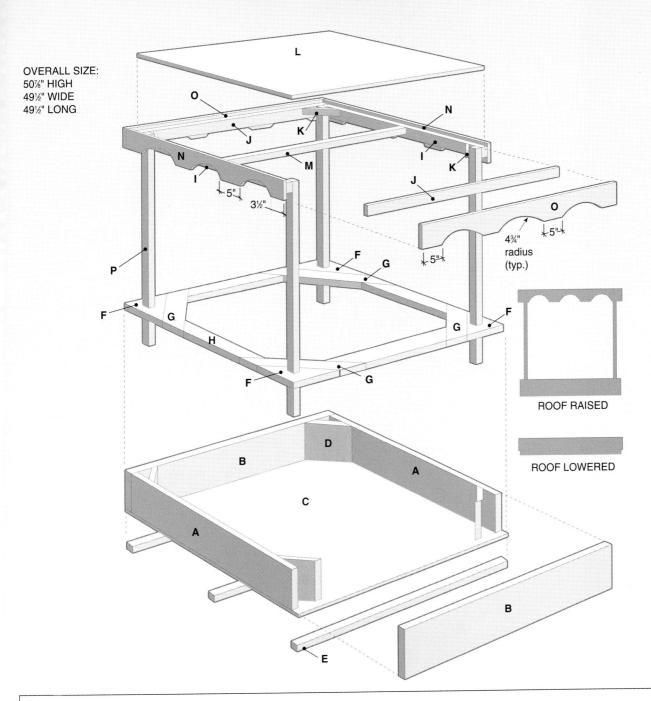

OVERALL SIZE:
50⅞" HIGH
49½" WIDE
49½" LONG

5"

3½"

4¾"
radius
(typ.)

5"

5"

ROOF RAISED

ROOF LOWERED

Cutting List				
Key	**Part**	**Dimension**	**Pcs.**	**Material**
A	End rail	1½ × 7¼ × 45"	2	Cedar
B	Side rail	1½ × 7¼ × 48"	2	Cedar
C	Bottom panel	⅜ × 48 × 48"	1	Plywood
D	Post cleat	1½ × 7¼ × 7"	4	Cedar
E	Base cleat	1½ × 1½ × 48"	3	Cedar
F	Corner brace	1½ × 7¼ × 7¼"	4	Cedar
G	Corner seat	1½ × 5½ × 21¼"	4	Cedar
H	Ledge	1½ × 3½ × 25"	4	Cedar

Cutting List				
Key	**Part**	**Dimension**	**Pcs.**	**Material**
I	End frame	1½ × 3½ × 45"	2	Cedar
J	Side frame	1½ × 3½ × 48"	2	Cedar
K	Top post cleat	¾ × 1½ × 5⅝"	4	Cedar
L	Top panel	⅜ × 48 × 48"	1	Plywood
M	Stringer	1½ × 1½ × 45"	1	Cedar
N	End apron	¾ × 5½ × 48"	2	Cedar
O	Side apron	¾ × 5½ × 49½"	2	Cedar
P	Post	1½ × 1½ × 48"	4	Cedar

Materials: Moisture-resistant wood glue, galvanized deck screws (1¼", 3"), 8d galvanized finish nails.

Note: Measurements reflect the actual size of dimensional lumber.

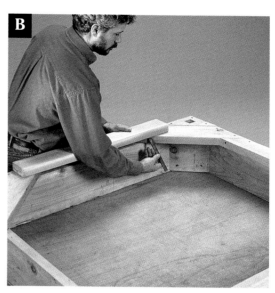

Install the corner braces to create post standards and to strengthen the corners of the sandbox.

Lay the ledges between the corners and trace cutting lines where they meet the corner seats.

Directions:
Covered Sandbox

BUILD THE SANDBOX FRAME. The sandbox frame is a simple box frame made from 2 × 8 cedar. Cut the end rails (A) and side rails (B) to length. Lay the rails on edge on a flat work-surface. Place the end rails between the side rails flush with the ends of the side rails to form a square frame. Drill three evenly spaced pilot holes for 3" deck screws at each joint, drilling through the side rails and into the end of the end rails (you may want to clamp the pieces together first). Countersink the pilot holes slightly so the screw heads will be recessed. Apply moisture-resistant glue to the boards at each joint, then drive a 3" deck screw at each pilot hole.

ATTACH THE BASE. The sandbox has a cedar plywood base that rests on three 2 × 2 cedar slats to minimize ground contact with the sandbox. Cut a 4 × 8 sheet of ⅜"-thick textured cedar siding (a plywood product) in half to make the bottom

panel (C). Cut carefully so you can use the other half of the sheet for the cover. Fasten the bottom panel, smooth surface up, to the bottom edges of the sandbox frame pieces with moisture-resistant glue and 1¼" deck screws. Cut the 2 × 2 base cleats (E), and fasten an outer cleat directly below each end rail (A), using glue and 3" deck screws driven through the cleat and into the base and end rail. Fasten the third cleat so it is centered between the outer cleats, flush with the ends of the frame. For this cleat, use glue and 1¼" deck screws driven through the inside surface of the bottom panel and into the top edge of the cleat. After the bottom panel and cleats are installed, cut and install post cleats (D). These parts have beveled edges so they can fit into each corner of the sandbox, creating pockets that will hold the roof posts. To cut the post cleats, set your circular saw to make a 45° cut, then trim the end of a cedar 2 × 8 so the bevel points in. Now, measure 7" from the top of the

bevel, and mark a cutting line. With your saw still set at 45°, cut at this line, so the bevel slants in toward the first bevel. Flip the 2 × 8 over, and continue cutting cleats, using the same technique. When all four cleats are cut, fasten them in each corner with glue and 1¼" deck screws, driven roughly perpendicular to the boards in the sandbox frame **(photo A).**

CUT & ATTACH THE TOP OF THE SANDBOX FRAME. The top part of the sandbox frame is made up of the corner boards and the ledges that create the seating area around the perimeter of the sandbox. Cutting these parts involves cutting some miters. The easiest, most accurate way to cut the top parts for the frame is by laying boards directly in place and using the other parts of the framework to trace cutting lines. Start by cutting the corner braces (F): cut two pieces of cedar 2 × 8 to 7¼" in length, then draw a diagonal cutting line on each board. Cut along the cutting line of each board to create four 7¼" triangular pieces.

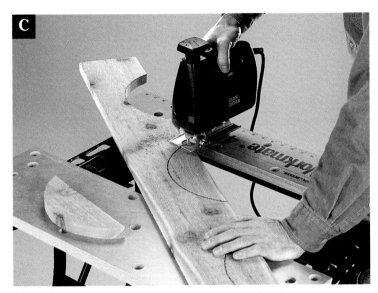

Make decorative cutouts on the roof aprons with a jig saw.

you made the post cleats for the base. Fasten the top post cleats with glue and screws. To make the top panel (L), use the other half of the cedar plywood panel that was cut for the base. Fasten it to the top frame with glue and 1¼" deck screws. Cut the end aprons (N) and side aprons (O) to length from 1 × 6 cedar. Cut 4¾"-radius semi-circles, spaced at 5" intervals, into the aprons to dress up the design. Cut the semi-circles with a jig saw **(photo C).** Attach the aprons to the end and side frames, so the tops are flush with the top panel, with glue and 1¼" screws.

APPLY FINISHING TOUCHES. Cut the posts (P) to length from 2 × 2 cedar. Test the fit of the posts: set a post in each corner-brace cutout; then, set the roof onto the tops of the posts **(photo D).** Sand all surfaces of the sandbox and cover smooth, then apply several coats of clear wood sealer. Move the sandbox to its planned location, then fill with sand.

Each corner brace has a 1½" square cutout for a 2 × 2 roof post. To mark positions for these cutouts, use a piece of 2× cedar to mark reference lines 1½" inside each of the short arms of the triangle. Then, set a cedar 2 × 2 on end at the corners where each set of reference lines joins. Trace around the 2 × 2 to make a cutout on each end brace. Drill a starter hole at each corner of the cutout, then cut all four cutouts with a jig saw. Fasten the corner braces in each corner of the frame using glue and 3" deck screws, countersunk slightly. Now, lay a piece of 2 × 6 cedar next to one of the corner braces, so it is resting flat on the frame and the edge is flush with the inside edge of the corner brace. From below, trace the outside edges of the end and side rails onto the underside of the 2 × 6 to mark cutting lines for a corner seat (G). Cut along the cutting lines with a circular saw, then mark and cut the three other corner seats. Install them so the edges are butted up against the corner

braces, using glue and 3" screws. Finally, cut the ledges (H) slightly longer than the finished length (see *Cutting List,* page 9) from 2 × 4 cedar. Lay each ledge in place, so it spans between two corners, with the outside edge flush with the outside of the sandbox. Mark cutting lines where ledges intersect the corner seats **(photo B);** cut the ledges with a circular saw, then install with glue and screws. Also drive two 8d galvanized finish nails through each joint for extra strength.

BUILD THE ROOF. The roof is made in much the same way as the sandbox frame. Cut the end frames (I) and side frames (J), then position the end frames between the side frames, and fasten with 3" deck screws and glue at each joint (drill countersunk pilot holes before driving screws). Cut the stringer (M) from 2 × 2 cedar and fasten it between the frame ends, halfway between sides, with glue and screws. Cut the top post cleats (K) to length from 1 × 2 cedar, cutting 45° bevels on both ends, the same way

With the ends of the posts in the corner brace cutouts, slip the roof over the top post ends.

Oak Desk

This little oak desk gives new meaning to the three R's:
Red oak, Right angles, and Really easy to build.

CONSTRUCTION MATERIALS

Quantity	Lumber
1	¼" × 2 × 4' hardboard
1	½ × 2" × 4' red oak
1	½ × 3" × 2' red oak
1	¾" × 2 × 4' oak plywood
1	1 × 2" × 8' red oak
1	1 × 3" × 4' red oak
2	1 × 4" × 8' red oak
2	2 × 2" × 8' red oak

Here's a nice alternative to the scores of plastic and particleboard novelty desks that are sold as childrens' furniture. Made of oak and oak plywood, this small-scale version of an old-style school desk makes a great work and study area for your youngster. The spacious desktop provides plenty of workspace for drawing and reading.

The roomy "childproof" drawer has a built-in stop so it can't be pulled all the way out.

The simplicity of this desk design makes it easy to build, but it also gives it a timeless beauty that will blend into just about any decorating scheme. And best of all, you can save hundreds of dollars over comparable furnishings you might buy at a store.

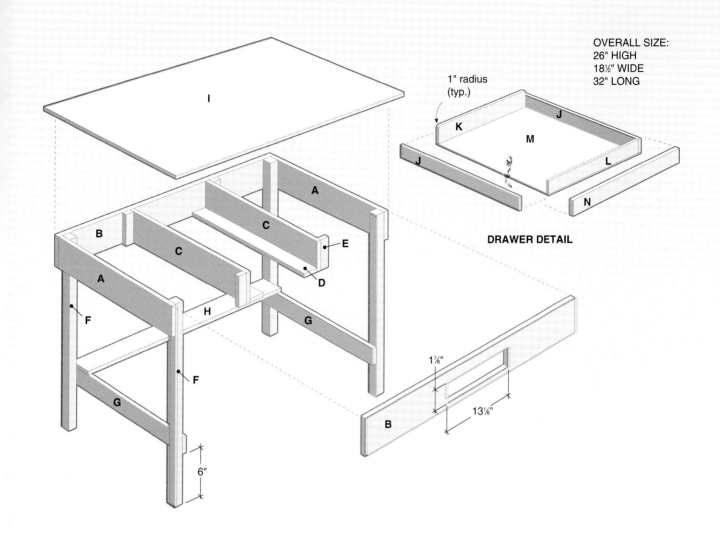

OVERALL SIZE:
26" HIGH
18½" WIDE
32" LONG

1" radius
(typ.)

DRAWER DETAIL

Cutting List

Key	Part	Dimension	Pcs.	Material
A	End rail	¾ × 3½ × 14½"	2	Oak
B	Side rail	¾ × 3½ × 30"	2	Oak
C	Slide side	¾ × 3½ × 14½"	2	Oak
D	Slide bottom	¾ × 1½ × 14½"	2	Oak
E	Slide cleat	¾ × 1½ × 3½"	4	Oak
F	Leg	1½ × 1½ × 25¼"	4	Oak
G	Leg stretcher	¾ × 3½ × 14½"	2	Oak

Cutting List

Key	Part	Dimension	Pcs.	Material
H	Stringer	¾ × 3½ × 25½"	1	Oak
I	Desktop	¾ × 19 × 33"	1	Oak plywood
J	Drawer side	½ × 1½ × 14"	2	Oak
K	Drawer back	½ × 2½ × 12"	1	Oak
L	Drawer front	½ × 1½ × 12"	1	Oak
M	Drawer bottom	¼ × 13 × 14"	1	Hardboard
N	Drawer faceplate	¾ × 2½ × 14"	1	Oak

Materials: Wood glue, brass wood screws (#6 x 1", #6 x 1¼", #6 x 1½"), brass corner brackets (8), oak wood plugs, oak drawer knob, wood putty, oak veneer edge tape (9'), finishing materials.

Note: Measurements reflect the actual size of dimensional lumber.

Cut out the drawer opening in the front side rail with a jig saw—drill starter holes at the corners.

Position the slides on the marks and secure in place with wood glue and screws driven through the cleats and into the side rails.

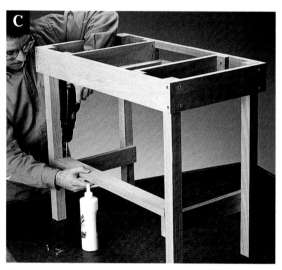

Attach the stringer to the leg stretchers with wood glue and countersunk screws.

Directions: Oak Desk

BUILD THE FRAME & DRAWER SLIDE. Cut the end rails (A) and side rails (B), and sand them smooth. Lay the side rails edge to edge and measure 8⁷⁄₁₆" in from each. Draw a line that spans both side rails. On one side rail (this will become the front rail), measure ⅞" down from the top edge and ¾" up from the bottom edge and draw lines to form a 13⅛ × 1⅞" rectangular cutout for the drawer. Drill a ⅜"-dia. starter

hole at each corner, and make the rectangular cutout with a jig saw **(photo A).** Smooth the edges with a sanding block. Cut the slide sides (C) and slide bottoms (D), and sand smooth. Position a slide bottom flush against the bottom of each slide side, and fasten with wood glue and #6 × 1½" brass wood screws. Drill pilot holes for the screws, countersunk just enough so the screw heads will be recessed. Cut the slide cleats (E) and fasten them to the ends of the slide sides with glue and countersunk screws. Position the slide assemblies so the inside faces of the slide sides are even with the sides of the drawer cutout and the matching lines on the back rail. The tops of the slide bottoms should be level with the bottom of the drawer cutout. Secure the slide assemblies with glue and countersunk screws driven through the cleats and into the side rails **(photo B).** Next, place the end rails between the side rails, flush with the ends, to form a rectangular frame. Clamp the frame, then

drill counterbored pilot holes (make sure the counterbores are sized to accept oak wood plugs) through the side rails and into the end rails. Unclamp the frame, apply glue to the joints, then reassemble it and drive #6 × 1½" brass wood screws into the pilot holes.

BUILD & ATTACH THE LEGS, STRETCHERS & STRINGERS. These support elements add stability to the desk. Cut the legs (F) to length from 2 × 2 oak and sand with medium-grit sandpaper. Turn the rail and slide assembly upside down on your worksurface. Clamp a leg in each corner, with the top of the leg flush with the top edge of the frame. Drill counterbored pilot holes through the end rails and into the legs. Unclamp the legs, apply glue to the joints, then secure the joints with screws driven into the pilot holes. Set the assembly in an upright position. Cut the leg stretchers (G) and stringer (H). Measure up from the bottom of each leg 6" and place a reference mark on the inside surface of each leg. At-

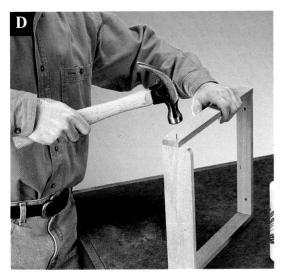

Assemble the drawer with glue and 3d finish nails. Be sure to drill pilot holes for the finish nails.

Trim the veneer edge tape with a sharp utility knife, then sand the desktop edges and surfaces.

tach the leg stretchers to the legs, just above the reference lines, so the ends of the stretchers are flush with the outside faces of the legs. Position the stringer on top of the stretchers, against the inside faces of the back legs. Attach the stringer to the leg stretchers with glue and counterbored screws **(photo C).**

BUILD THE DRAWER. The drawer construction is simply a box with an attached faceplate. We used ½"-thick oak for the drawer frame—you can find ½" oak with the shelving and molding at most building centers. Cut the drawer sides (J), drawer front (L), drawer back (K) and drawer bottom (M). Draw roundover cutting lines with a 1" radius at the top corners of the drawer back, and cut with a jig saw. Position the drawer front and back between the drawer sides, flush with the ends of the drawer sides. Assemble the drawer with glue and 3d finish nails **(photo D).** Be sure to drill pilot holes for the finish nails. The pilot holes should be thinner than the finish nails. Place the drawer frame upside down, and attach

the drawer bottom with glue and 3d finish nails. Turn the drawer unit over and place it on two pieces of ½"-thick scrap wood. Cut the drawer faceplate (N) to length from 1 × 3 oak. Place the faceplate upright on edge against the drawer front so it overhangs the drawer front by ½" on all sides. Fasten the faceplate to the drawer front with #6 × 1" brass wood screws, driven through countersunk pilot holes in the inside face of the drawer front, and into the back face of the faceplate. Tip the front of the drawer up, and insert the back edge into the drawer opening in the front rail. Test the drawer.

BUILD & ATTACH THE DESK-TOP. Cut the desktop (I) to size from ¾"-thick oak plywood. Sand the edges and surfaces with medium-grit sandpaper. Apply self-adhesive wood veneer edge tape to the edges by pressing with a household iron (the heat activates the glue). Trim any overhanging edges with a utility knife **(photo E).** Fasten the desktop to the desk with a brass corner bracket (or L-brace) at each side of all four

corners. The desk top should overhang all sides by 1½" **(photo F).**

APPLY FINISHING TOUCHES. Fill nails holes with oak-tinted wood putty, and glue oak plugs into screw counterbores. Sand the plugs so they are flush with the surface, then finish-sand all surfaces with 180-grit sandpaper. Apply stain (we used light oak wipe-on stain) and topcoat (we used three coats of polyurethane on the desk top, and two coats on the rest of the desk).

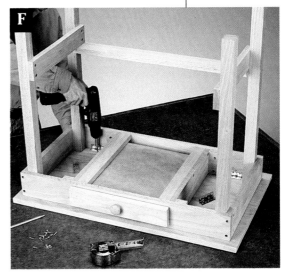

Fasten the desktop to the side rails with corner brackets, leaving a 1½" overhang on all four sides.

Playhouse

New on market: spacious 1 rm playhouse, cheerful decor, many windows, folds up for storage. Take immediate possession.

CONSTRUCTION MATERIALS

Quantity	Lumber
4	¼" × 4 × 8' plywood
30	1" × 2 × 8' premium furring
3	1" × 4 × 8' pine

Kids of all ages long for a place of their own: a tree house, a backyard fort; a seldom-used shed; anything with four walls, a roof and a door. The problem is, most of these would-be retreats are either eyesores or they are well outside of a parent's field of supervision. By building this charming little playhouse with walls that have more window than wood, you can give a child the private space that he or she longs for, but still keep the little one within your sight.

Made from lightweight ¼" plywood, this playhouse is easy to move around and store. The walls are hinged together and will fold flat against a wall when the lift-off roof is removed. You can even set the playhouse up in the backyard on a pleasant afternoon to give your child a room with a view.

OVERALL SIZE:
65" HIGH
35" WIDE
72" LONG

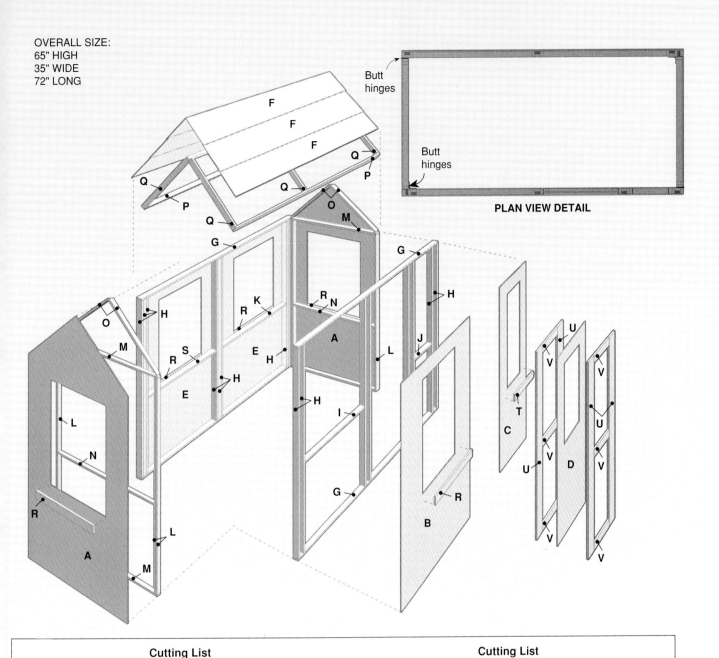

PLAN VIEW DETAIL

Cutting List						Cutting List				
Key	**Part**	**Dimension**	**Pcs.**	**Material**		**Key**	**Part**	**Dimension**	**Pcs.**	**Material**
A	House end	¼ × 31½ × 59⅞"	2	Plywood		L	Corner stud	¾ × 1½ × 47¼"	6	Pine
B	House front	¼ × 36 × 48"	1	Plywood		M	End plate	¾ × 1½ × 31½"	4	Pine
C	House side	¼ × 16 × 48"	1	Plywood		N	End sill	¾ × 1½ × 29¼"	2	Pine
D	Door	¼ × 19½ × 46"	1	Plywood		O	Gable frame	¾ × 1½ × 18¼"	4	Pine
E	Back wall	¼ × 48 × 72"	1	Plywood		P	Roof frame	¾ × 1½ × 80"	4	Pine
F	Roof panel	¼ × 7⅞ × 80"	6	Plywood		Q	Roof frame	¾ × 1½ × 19¾"	6	Pine
G	Side plate	¾ × 1½ × 72"	4	Pine		R	Window ledger	¾ × 1½ × 22"	5	Pine
H	Wall stud	¾ × 1½ × 46½"	29	Pine		S	Back wall sill	¾ × 1½ × 30"	1	Pine
I	Front sill	¾ × 1½ × 28½"	1	Pine		T	Window ledger	¾ × 1½ × 12"	1	Pine
J	Side sill	¾ × 1½ × 10"	1	Pine		U	Door stile	¾ × 3½ × 46"	4	Pine
K	Back wall sill	¾ × 1½ × 31½"	1	Pine		V	Door rail	¾ × 3½ × 19½"	6	Pine

Materials: Glue, wood screws (#6 × 1", #6 × 1½", #6 × 2"), 4d finish nails, 3 × 3" butt hinges (14), 1½ × 3" butt hinges (3), finishing materials.

Note: Measurements reflect the actual size of dimensional lumber.

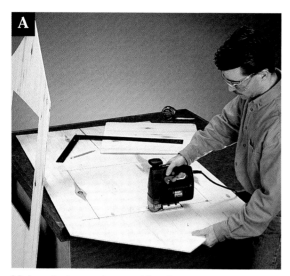
Make window cutouts with a jig saw.

Mark the 1 × 2 wall stud positions on the side plates with a combination square and a pencil.

Directions: Playhouse

MAKE THE PANELS. The panels are lightweight skins that are fastened to the structural framework of the playhouse. Begin construction by cutting the house end panels (A), house front panel (B), house side panel (C), door (D) and back wall panel (E). To economize on material, lay out cutting guidelines carefully before cutting the parts. Sand all the parts smooth after cutting. To cut the angles on the end panels, measure 15¾" across the top end of the panel and place a mark. Measure 48¾" up from the bottom end on the sides and mark points. Use a straightedge to connect the points to the centerpoint on the top end. Cut along the diagonal lines with a jig saw to form the gable profile on the tops of the end panels. To make the window cutouts on the end panels, draw a line from side to side, 24" up from the bottom. Mark another line 45" up from the bottom. Mark points along

each line, 5¾" in from each side. Draw guidelines connecting the points to form a rectangle. Drill pilot holes in the corner of the layout and cut along the guidelines with a jig saw to form the 20"-wide × 21"-high window cutouts on the end panels **(photo A).** Make similar window cutouts on the house front panel, house side panel and back wall panel by finding the center of each panel and drawing lines 24" and 45" up from their bottom edges. On the front panel, make the sides of the window cutout 10" on each side of center. On the side panel, make the side of the window cutout 5" on each side of center. Unlike the other panels, the back wall panel has two cutouts, so mark centerpoints 18" from each end, and make the sides of the cutouts 10" to each side of those centerpoints. On the door, draw window cutout lines 23" and 44" up from the bottom, and mark their centerpoints. Mark points 6½" to each side of the centerpoints to form the cutout shape. Sand the edges smooth.

MAKE THE SIDE PLATES & WALL STUDS. The frames at the front and back of the playhouse are made by attaching ¾"-thick wall studs between two pairs of side plates. Building this part of the playhouse is very similar to frame carpentry, if you've ever built a stud wall before. The frame for the front panel and the frame for the back wall panel are the same, except for the door and small window on the front. Start by cutting the side plates (G) and wall studs (H) to size. Lay the side plates in pairs on a flat worksurface. Before attaching the studs between the side plates, you must carefully lay out the stud positions. Hook your measuring tape on one end of a side plate. Mark lines 1½", 3¾", 33", 35¼", 56", 58¼" and 69" from one end on the face of the side plate. These marks show the post positions on the front side plates, and are in order, left to right, if you are facing the front of the finished playhouse. Using a combination square, draw lines across the plates, ¾" to the right of each mark, showing the

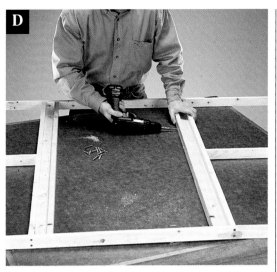

Drive countersunk wood screws through the frame sections into the door-frame studs.

Attach a stud to each side of the door opening to create a solid door frame.

position of each ¾"-thick stud **(photo B).** To mark the stud positions on the wall side plates, measure and mark lines at 2¼", 33", 35¼", 67½" and 69¾" from one end of the side plate. These markings are in sequence, left to right, if you are facing the front of the playhouse. At each end of the side plate pairs, use glue and counterbored #6 × 2" wood screws to attach a stud between the side plates.

MAKE THE FRONT FRAME. The front frame is made up of wall studs and sills that fit horizontally between the studs, just behind the bottom edges of each window cutout. These sill and stud assemblies are attached between side plate pairs with glue and #6 × 2" wood screws. Cut the back wall sills (K, S), front sill (I) and side sill (J) to size from 1 × 2 pine. Use glue and counterbored wood screws to fasten the front sill and side sill between the stud pairs, 22½" up from the bottoms of the studs—these frame parts fit behind the front panel. Position the front sill and its studs between the side plates, start-

ing at the second stud position lines on the left-hand side of the side plates (drawn previously on the side plate faces). Fasten the studs with glue and counterbored wood screws, driven through the side plates and into the ends of the studs. Insert a stud in the gap between the front sill assembly and the stud at the end of the frame. Make sure the inserted stud butts against the front sill assembly—the face should be flush with the inside edges of the side plates. Fasten it with glue and counterbored wood screws. Position the side sill and its stud pair on the right-hand side of the front frame, making sure the studs are aligned with their designated position lines. Attach these studs with glue and counterbored wood screws. Insert a stud in the gap between the side sill assembly and the stud on the end of the frame. (Unlike the inserted stud on the left side of the frame, this stud will completely fill the gap.) Make sure the inside face of the inserted stud is flush with the inside edges of the side

plates, and fasten it with glue and wood screws. Now, make the door frame. To make a secure anchor for the door and its hinges, you need to attach two studs on either side of the opening that was formed when you attached the side sill assembly and front sill assembly. Butt the studs against the assemblies, edge first, and fasten the studs so their faces are flush with the inside edges of the wall. Attach the parts with glue and counterbored wood screws, driven through one stud face and into the other stud side **(photo C).** Complete the door frame by attaching the two remaining studs, face first, to the door frame. Make sure the inside edges are flush, and attach the parts with glue and counterbored #6 × 2" wood screws **(photo D).**

MAKE THE WALL FRAME. The wall frame is similar to the front frame, but it has no door space. Start by attaching the wall sill between two studs with glue and counterbored wood screws, 22½" from the stud bottoms. Fasten a back wall sill (K) between two studs at the same height. Fasten the wall sill and studs between the plates, starting at the second stud position line on the right-hand side of the side plates (drawn previously on the side plate faces). Insert a stud in the gap between the wall sill assembly and the stud at the end of the frame. Make sure the inserted stud butts against the wall sill assembly—the face should be flush with the inside edges of the side plates. Fasten it with glue and counterbored wood screws. Fasten the second back wall sill (S) and studs between the side plates, starting at the first stud position lines on the left-hand side of the side plates. Insert a stud in the gap between the shorter wall sill assembly and the stud at the end of the frame. Attach it with glue and wood screws. Fasten a stud in the gap between the two frame assemblies at the center of the wall. Position this stud so the inside face is flush with the inside edges of the side plates. Use glue and counterbored wood screws to attach the stud.

MAKE THE END FRAMES & ROOF FRAMES. Start by cutting the corner studs (L) and end plates (M) to size. Use glue and counterbored wood screws to attach two corner studs between the two end plates. Position the corner studs so their outside faces are flush with the

Attach the roof frames with butt hinges.

Install a nonlocking doorknob set in the door.

end plate ends, and drive screws through the end plates into the corner stud ends, forming two rectangular frames. Cut the end sills (N) to size, and attach them to the remaining corner studs with glue and wood screws. The end sills should be 22½" from the corner stud bottoms. Position the attached end sills and studs into each rectangular end frame, and attach

them with glue and wood screws. To make the roof frames, cut the roof frame boards (P, Q) to size. Use glue and counterbored wood screws to attach the shorter roof frame boards between the longer roof frame pairs at each end and at their centers. When both sections are assembled, attach them with evenly spaced, 3 × 3" butt hinges **(photo E).**

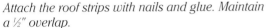

Attach the roof strips with nails and glue. Maintain a ½" overlap.

Position the finished roof onto the playhouse.

ATTACH THE PANELS & FRAMES. Attach the house end panels to the end frames with glue and counterbored wood screws, driven through the end panels and into the end frames. Attach the panels to the end frames so that one end frame has the double-end stud at the front, and the other at the back—when the frames are attached this will be very important for hinge placement. Make sure the bottom and side edges are flush. Cut the gable frames (O) to size, and use glue and wood screws to attach them along the top of the panels, flush with the top edges. Use glue and wood screws to attach the back wall panel to the wall frame, making sure the window cutouts are unobstructed by the frame and the edges are flush. Attach the front and side panels to the front frame. Cut the door stiles (U) and door rails (V) to size. Glue the door stiles, and sandwich the door panel between stile pairs on the panel edges. Drive wood screws through the inside stiles and into the outside stiles. Attach the door rails along the top and bottom edges of the door panel in the same manner. Attach the final door rails on the bottom edge of the door panel's window cutout. Drill the holes for the doorknob and latch, and hang the door in the frame with $1\frac{1}{2} \times 3$" butt hinges. Install non-locking doorknob hardware **(photo F).**

ATTACH THE FRAMES. The front, wall and end frames are connected to each other on both sides with two evenly spaced 3×3" butt hinges. To ensure effective fold-up of the playhouse, the hinge barrels on the end frame sides with two end posts should face out. The hinge barrels on the single end post sides should be attached facing in (see *Diagram*, page 17).

MAKE THE ROOF & LEDGERS. Cut the roof panels (F) to size. Fold the roof frames in half, and fasten the roof panels to the frames with 4d finish nails, starting at the bottom edges.

(Keep the bottom edge flush with the frame.) Overlap the next roof panel by ½" on top of the first **(photo G),** and again with the final roof panel. Turn the roof over, and repeat the procedure with the remaining roof frame. The top panels will overhang the roof peak by about 1½". Set all the nails, and position the roof on the playhouse **(photo H).** Cut the window ledgers (R, T) to size. Center them on the outside, bottom edge of each window cutout, and fasten them with glue and wood screws to strengthen the window edges.

APPLY FINISHING TOUCHES. Fill screw holes with wood putty, then sand all surfaces. Cover all the surfaces with a washable, semi-gloss enamel paint. We used a fairly fancy paint job with our playhouse. We painted the walls one color, then painted the trim a contrasting color. We even chose to paint the roof and door another color.

Bathroom Stool

Perfectly sized for children as they put on their shoes or stand at the sink to brush their teeth, this stool makes it easy to be small.

Well suited for a bathroom or a bedroom, this versatile stool can give your child just the lift he or she needs. Made from solid red oak, it features a sturdy four-legged construction that won't wobble or tip. It is low enough to the ground that a typical toddler can use it as a dressing seat when struggling to put on shoes and socks. It is just high enough to give the added boost he or she needs to reach the sink. Because it is lightweight and portable, it can even move around the house with your child to help with a wide array of activities and chores.

The warm tones of the solid oak and the simple lines of the design make this bathroom stool an attractive accent in any room. But part of its beauty comes from the ease with which it can be built. The pedestal-style base is made with four basic 1 × 6 legs butted against a central post that is made from four 1 × 2s. The round seat is simply made from two semicircular pieces of oak 1 × 6, butted together along their straight edges. Even the cleats that fit under the seat are easy to cut with a power miter box or a miter box and backsaw.

CONSTRUCTION MATERIALS

Quantity	Lumber
1	1 × 2" × 4' red oak
1	1 × 6" × 6' red oak

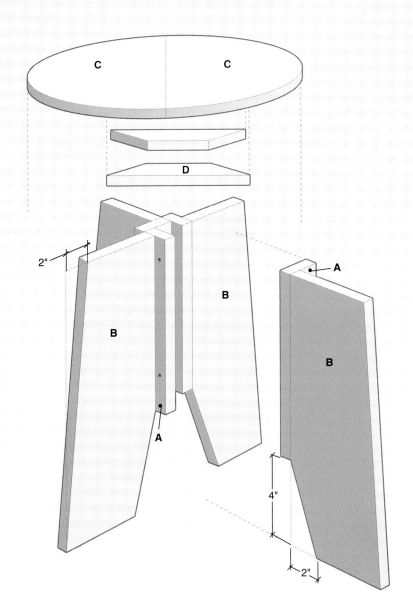

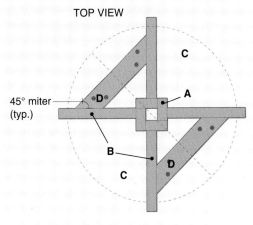

TOP VIEW

45° miter
(typ.)

C

D

A

B

C

D

SIDE VIEW

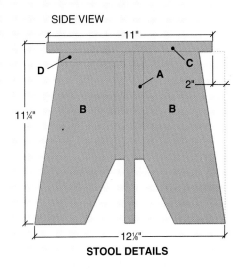

11"

D

C

A

2"

11¼"

B

B

12⅛"

STOOL DETAILS

Cutting List				
Key	**Part**	**Dimension**	**Pcs.**	**Material**
A	Post side	¾ × 1½ × 6½"	4	Red oak
B	Leg	¾ × 5½ × 10½"	4	Red oak
C	Seat section	¾ × 5½ × 11"	2	Red oak
D	Cleat	¾ × 1½ × 6"	2	Red oak

Materials: Glue, brass wood screws (#6 × 1¼", #6 × 2"), ⅜"-dia. oak wood plugs, finishing materials.

Specialty Items: Power miter box (optional), compass.

Note: Measurements reflect the actual size of dimensional lumber.

Directions:
Bathroom Stool

MAKE THE LEGS. The four legs for the stools are cut from 1 × 6 oak. The legs are attached to a central post made from four pieces of oak 1 × 2. Cut the legs (B) to size. Each leg is trimmed on the sides, removing two triangular strips to create a tapered appearance. Draw cutting lines onto each leg **(photo A).** To make the lines, mark points 2" in from the inside, bottom corner of each leg, on the bottom edges. Mark a point 4" up from the same corner, on the inside edge of each leg. Connect the points to make cutting lines. To mark the triangular cutouts on the outside edges of the legs, mark points 2" in from the top, outside corners, on the top edge of each leg. Connect each point to the outside, bottom corner of the leg. Cut out the two triangular shapes from each leg, using a jig saw or circular saw.

MAKE THE CENTRAL POST. Cut the four post sides (A) to length. Use a straightedge to draw reference lines ¾" in from one side edge of each post (these lines mark the inside edge of the adjoining post where it butts against each post). Drill two pilot holes for #6 × 2" wood screws between each reference line and the edge of each post— drill the pilot holes 2" down from the top and 2" up from the bottom. Counterbore each pilot hole to accept a ⅜"-dia. oak wood plug—the counterbored faces will be the outside faces of

TIP

Red oak is the most commonly available type of oak lumber. It is relatively hard, so make sure you drill pilot holes for every screw or nail you use. And because steel can cause oak to discolor, always use brass fasteners.

Draw cutting lines on both edges of each leg to create a tapered appearance. Cut with a jig saw or circular saw.

the post sides. Also drill two plain pilot holes on the back of each post side, ⅜" in from the opposite edge of the counterbored pilot holes (these pilot holes are for the screws you will use to connect each post side to a leg). Drill 1" from the top and bottom edges. Apply glue to the inside edge of one leg, and butt a post side against the leg, so the tops of the pieces are flush. Extend the pilot holes at the inside edge of each post into the leg, then fasten the leg to the post with #6 × 2" brass wood screws—always use brass fasteners when working with oak. Fasten each leg to a post side.

ASSEMBLE THE PEDESTAL. To assemble the stool pedestal, apply glue to the edge of each post side that is flush with a leg. Position a pair of leg/post sides together so the tops are flush and the counterbored pilot holes in the post sides are centered on the glued edge of the adjoining assembly. Drive #6 × 1¼" brass wood screws through

the pilot holes in the posts and into the edge of the adjoining post. Repeat these steps with the rest of the leg assemblies to complete the pedestal **(photo B).**

MAKE THE SEAT SECTIONS. The seat is made from two semicircular boards that are butted together to form a full circle. Cut a piece of 1 × 6 oak into two 11"-long seat sections (C). Lay the seat sections side by side. Use a compass to draw a circle with a 5½" radius on the seat sections. The point of the compass should be positioned between the boards, centered end to end. Cut along the circle with a jig saw to form two semicircles. Attach a belt sander to your worksurface so the belt is perpendicular to the surface, and use it to smooth out the rough spots and contours on the seat sections **(photo C).**

MAKE & ATTACH CLEATS. The seat sections are edge-glued together, with the joint reinforced by a pair of wood cleats

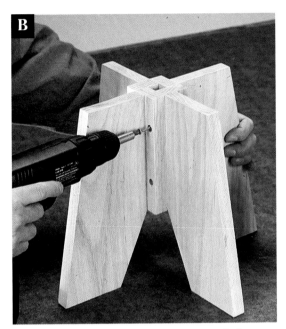

B

Connect the posts with wood screws and glue to complete the base.

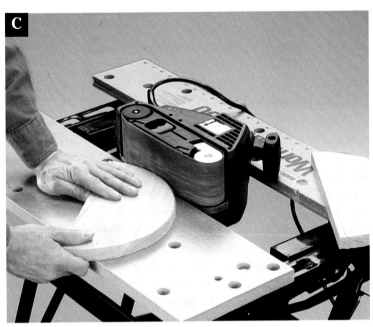

C

Smooth out the curves on the seat sections with a belt sander after you cut them to shape.

that fit between the legs at the top of the pedestal, and span the seam between the seat sections. Cut the cleats (D) to length from 1 × 2 oak. Trim off the ends of each cleat at a 45° angle with a miter box (see *Diagram*, page 23). Drill a pair of staggered, counterbored pilot holes near each end of each cleat (see *Diagram*) for fastening the cleats to the seat sections. Position the cleats so they fit between pairs of legs, flush with the tops, on opposite sides of the pedestal. Drill counterbored pilot holes and attach the cleats between the legs with #6 × 1¼" brass wood screws— make sure the tops of the cleats are flush with the tops of the legs and central post.

ATTACH THE SEAT. Apply glue to the straight edges of the seat sections, then butt them together to form a circle. Tape the seam with masking tape on the top, then lay the seat top down on a flat worksurface. Invert the pedestal, then set it on

top of the underside of the seat. Make sure the seat overhang is equal on all sides, and the seam of the seat falls across the centers of the cleats. Drive #6 × 1¼" screws through the cleats and into the seat. Drill counterbored pilot holes, and attach the seat to the top edges of the legs with one #6 × 2" screw driven through the seat and into the top of each leg.

APPLY FINISHING TOUCHES. Fill the counterbored screw holes in the seat with glued oak plugs **(photo D),** then sand the plugs level with the wood surface. Finish-sand all surfaces with 180-grit sandpaper. Apply wood stain (we used light oak wipe-on stain). After the stain has dried, apply two or three coats of polyurethane to protect the wood.

D

Insert glued oak plugs into the counterbored screw holes on the top of the seat.

Drawing Board

*Give your kids a good alternative to wall graffiti
with this erasable drawing board.*

CONSTRUCTION MATERIALS

Quantity	Lumber
6	1 × 2" × 8' pine
3	1 × 4" × 8' pine
1	1 × 6" × 4' pine
1	½" × 4 × 4' AB plywood
1	½ × 1¼" × 4' stop molding
1	¼" × 2 × 4' white melamine
1	½ × 12 × 36" cork sheet/tile

I t's an unexplained fact of nature that kids love to draw on walls. With this combination drawing board and bulletin board, you can confine your childrens' artistic inclinations to a suitable spot—and have a place to display drawings and artwork to boot.

The slick melamine surface on this drawing board erases easily and cleanly when common dry-erase markers are used. Nontoxic and available in an array of colors, dry-erase markers are sold at any office or school supply store. The cork used to make the bulletin board can be purchased as a sheet or in tile form from art or craft supply stores, or from just about any building center.

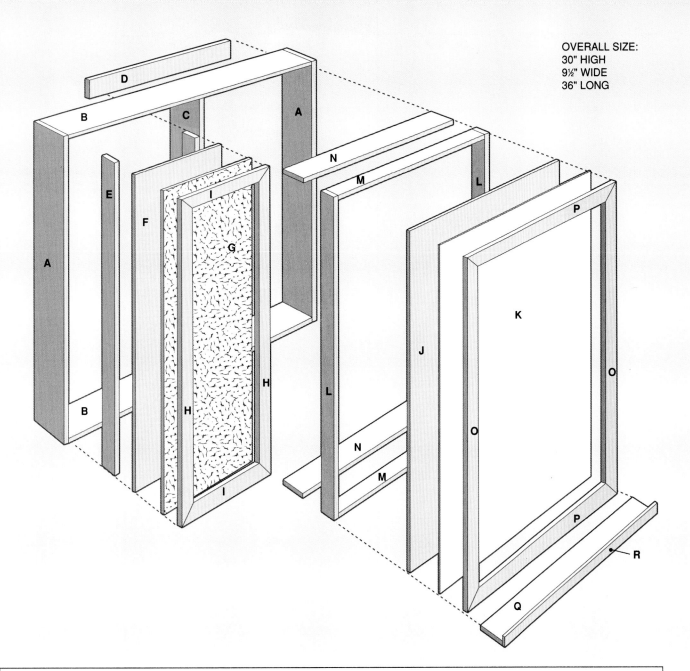

OVERALL SIZE:
30" HIGH
9½" WIDE
36" LONG

Cutting List				
Key	Part	Dimension	Pcs.	Material
A	Frame side	¾ × 3½ × 30"	2	Pine
B	Frame end	¾ × 3½ × 34½"	2	Pine
C	Stringer	¾ × 3½ × 28½"	1	Pine
D	Ledger	¾ × 3½ × 23¼"	1	Pine
E	Backer cleat	¾ × 1½ × 28½"	2	Pine
F	Cork backer	½ × 10½ × 28½"	1	Plywood
G	Cork board	½ × 10½ × 28½"	1	Cork tile/strip
H	Cork retainer side	¾ × 1½ × 30"	2	Pine
I	Cork retainer end	¾ × 1½ × 12"	2	Pine

Cutting List				
Key	Part	Dimension	Pcs.	Material
J	Board backer	½ × 22½ × 28½"	1	Pine
K	Marking board	¼ × 22½ × 28½"	1	Melamine
L	Board frame side	¾ × 1½ × 30"	2	Pine
M	Board frame end	¾ × 1½ × 22½"	2	Pine
N	Board cleat	¾ × 3½ × 22½"	2	Pine
O	Board retainer side	¾ × 1½ × 30"	2	Pine
P	Board retainer end	¾ × 1½ × 24"	2	Pine
Q	Marker tray	¾ × 5½ × 24"	1	Pine
R	Tray lip	½ × 1¼ × 24"	1	Pine

Materials: Wood glue, #6 × 1½" and #8 × 3" wood screws, finish nails (4d, 6d), wood plugs, pine-tinted wood putty, and finishing materials.

Note: Measurements reflect the actual size of dimensional lumber.

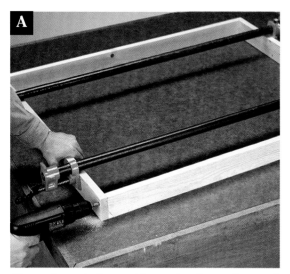

Clamp the main frame sides and ends together, and drill counterbored pilot holes at the joints.

Position the ledger between the stringer and right frame side and secure in place with wood glue and screws.

Directions:
Drawing Board

MAKE THE MAIN FRAME. The main frame for this drawing board houses the smaller frames that wrap the marker board and the bulletin board. Cut the frame sides (A) and frame ends (B) to length from 1 × 4 pine. Position the frame ends between the frame sides, to create a box frame. Use bar or pipe clamps to hold the frame together while you drill pilot holes through the sides and into the frame ends. Counterbore the pilot holes deeply enough to accept ⅜"-dia. wood plugs **(photo A).** Unclamp the frame, then apply wood glue to the ends of the frame ends and the mating surfaces of the frame sides. Fasten the frame components together with #6 × 1½" wood screws. Cut the stringer (C) and ledger (D) to length from 1 × 4 pine. Attach the stringer between the frame ends, so it is 12" from the left side of the frame. Use glue and counterbored screws. Next, position the ledger between the stringer and the right frame

side, flush with the back edge of the stringer and butted up to the inside surface of the top frame end **(photo B).** The ledger is used as an anchoring surface for mounting the drawing board to a wall. Secure the ledger with glue and screws driven through the frame side and the stringer. On the bulletin board side of the main frame (the left side), draw reference lines 1" in from the front edges of the left frame side and the stringer. The reference lines are used as guides for installing the backer cleats for the bulletin board. The 1" recess allows room for the backer and the cork panel to fit

TIP

An alternative to using melamine for a dry-erase writing board is to use a chalkboard panel. Chalkboard traditionally was made from solid slate. But these days, it is more commonly sold as a specially coated particleboard available in sheet form in thicknesses from ¼" to ½".

flush with the front edges of the frame opening. Cut the backer cleats (E) to length, then attach the cleats to the frame, just inside the reference lines, with glue and screws.

MAKE THE BULLETIN BOARD. The bulletin board is simply a plywood backer board and a ½"-thick cork panel, held in place by a mitered frame made from 1 × 2" pine. Build the bulletin board by cutting the cork backer (F) to size from ½"-thick plywood and fastening it to the backer cleats on the frame with glue and screws. Next, cut cork surface (G). You can use either strips of ¼"- to ½"- thick cork, or cork tiles. Position the cork material onto the backer board **(photo C)**—use panel adhesive to hold it in place if necessary. Next, make the retainer-strip frame that holds the cork surface and the backer board in place. Cut the cork retainer sides (H) and ends (I) to length from 1 × 2, mitering the corners at 45° to create a frame with square corners. Fasten the retainer frame pieces to the main frame side using glue and 6d finish nails. Be sure to drill

Cut the cork tiles or strips to size, and position them on top of the cork backer board.

Drive a 6d finish nail through each miter joint to lock it together—be sure to drill pilot holes.

pilot holes for the finish nails to prevent splitting the wood. After the retainer frame is built and attached to the main frame, drive one 6d finish nail through a pilot hole in each joint to lock-nail the mitered corners together **(photo D).**

MAKE THE MARKER BOARD. The marker board is similar in construction to the bulletin board, except that the marker board has a full frame both behind and in front of the primary surface. Start the marker board construction by cutting the board frame sides (L) and board frame ends (M) to length from 1 × 2 pine. Fasten the ends between the sides with glue and screws. Be sure to drill counterbored pilot holes for all screws. Draw a reference line ¾" in from the front edges of each board frame end. Cut the board cleats (N) to length and fasten them at the back sides of the reference lines with glue and screws. Cut the board backer (J) to size from ½"-thick plywood and fasten it to the board cleats with glue and screws. Cut the marking board (K) to size from ¼"-thick white

melamine and place it against the board backer. Cut the board retainer sides (O) and ends (P) to size to make a retainer frame for the marking board, following the same steps as for the cork retainer frame. Fasten the retaining frame to the board frame with glue and 6d nails.

MAKE & ATTACH THE MARKER TRAY. The marker tray is designed to hold dry-erase markers and erasers for quick and easy access. It has a lip in front to keep utensils and supplies from falling onto the floor. Cut the marker tray (Q) to length from 1 × 6 pine and cut the tray lip (R) to length from ½"-thick × 1¼"-wide stop molding. Attach the tray lip to the marker tray with glue and 4d finish nails. Fasten the marker shelf to the underside of the main frame, flush at the ends and back edge of the frame, using glue and countersunk #6 × 1½" screws **(photo E).**

APPLY FINISHING TOUCHES. Finish-sand the entire project and fill all counterbore holes with pine wood plugs, and use pine-tinted wood putty as

needed to fill any other holes or scratches. Apply your finish of choice. We used plain orange shellac (which both colors and protects the wood). When finished, hang the drawing board on your wall by driving #8 × 3" screws through the ledger and into wall stud locations. Check the drawing board with a level before driving the screws. Insert the marker board into the opening in the main frame, and drive screws through the main frame sides and into the marker board sides, fastening the drawing board together.

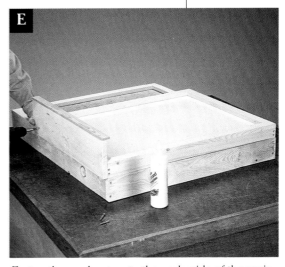

Fasten the marker tray to the underside of the main frame, using glue and screws.

Oak Table & Chairs

*Build a durable, attractive table-and-chair set just for the kids,
and watch them enjoy it for years to come.*

CONSTRUCTION MATERIALS

Quantity	Lumber
6	1 × 2" × 8' oak
2	1 × 3" × 8' oak
1	1 × 10" × 4' oak
1	¾" × 4 × 4' oak plywood
2	¾ × ¾" × 8' corner molding

A table-and-chair set just for kids, this project is great for games, puzzles or snack time. The tabletop features a convenient rim around the edges, which curbs messy spills and keeps all the pieces of a jigsaw puzzle or board game off the floor. The chairs are very nearly tip-proof and blend nicely with the table style. Though this project is made for children, the attractive styling and beautiful oak construction make it suitable in any decor. There is no doubt about the sturdiness of this solid oak project—it's truly built to last. In fact, it just might become a cherished family heirloom that is handed down from generation to generation.

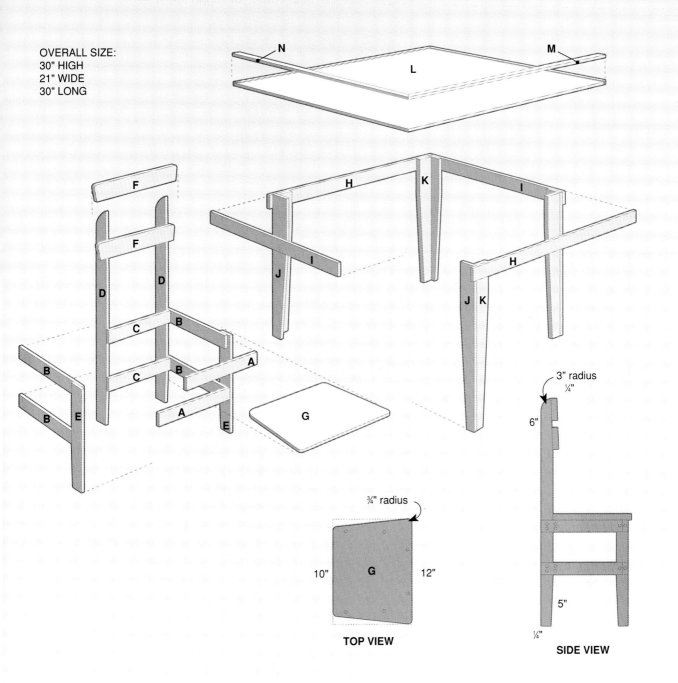

OVERALL SIZE:
30" HIGH
21" WIDE
30" LONG

3" radius
¼"

6"

5"

¼"

¾" radius

10"   G   12"

TOP VIEW

SIDE VIEW

Cutting List				
Key	Part	Dimension	Pcs.	Material
A	Front	¾ × 1½ × 9½"	4	Oak
B	Side	¾ × 1½ × 9¾"	8	Oak
C	Cross rail	¾ × 1½ × 8"	4	Oak
D	Post	¾ × 1½ × 25¼"	4	Oak
E	Leg	¾ × 1½ × 12"	4	Oak
F	Slats	¾ × 2½ × 12"	4	Oak
G	Seat	¾ × 9¼ × 12"	2	Oak

Cutting List				
Key	Part	Dimension	Pcs.	Material
H	Apron side	¾ × 1½ × 32"	2	Oak
I	Apron end	¾ × 1½ × 23½"	2	Oak
J	Outside leg	¾ × 2½ × 21¼"	4	Oak
K	Inside leg	¾ × 1½ × 21¼"	4	Oak
L	Top	¾ × 27½ × 36"	1	Oak plywood
M	End nosing	¾ × ¾ × 28"	2	Oak molding
N	Side nosing	¾ × ¾ × 36½"	2	Oak molding

Materials: Glue, brass wood screws (#6 × 1¼"), wire brads, finishing materials.

Note: Measurements reflect the actual size of dimensional lumber.

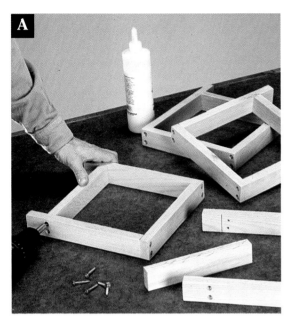

Attach the cross rails between the frame sides with glue and screws.

Sand the tapers into the posts and legs with a belt sander mounted sideways on your worksurface.

Directions: Oak Table & Chairs

BUILD THE CHAIR FRAMES. The *Cutting List* on page 31 shows the material needed for two small chairs, but you can build as many as you want or need with the directions provided. Start with the chair frames. Each chair contains two frames, which are made from four pieces of ¾"-thick oak. When the chair is completed, the top frame is located directly under the seat. The remaining frame is positioned 7" up from the bottoms of the legs. The entire chair assembly is built around the top frame, which serves as a kind of structural anchor for the seat and legs. Start by cutting the fronts (A), sides (B) and cross rails (C) to size. Sand all parts with medium-grit sandpaper to smooth out any rough edges after cutting. Drill a pair of counterbored pilot holes for #6 × 1¼" screws on the front face of each front, centered ⅜"

from each end. Drill pilot hole pairs for #6 × 1¼" wood screws on each side, centered ¾" and 1⅞" from one end of the sides. Fasten the fronts to the sides with glue and wood screws, driven through the pilot holes in the fronts and into the side ends—position the sides so that the ends with the pilot holes are facing away from the front. The top and side edges of the front and sides should be flush. Next, draw reference lines across the inside faces of each side, 1½" from the ends with the pilot holes. Glue the cross rail ends, and position them between the sides. Make sure the cross rails are lined up with the reference lines, and drive #6 × 1¼" wood screws through the pilot holes into the cross rail ends to secure them **(photo A).** Use sandpaper to smooth out the top and bottom edges of the frames.

BUILD THE POSTS & LEGS. The posts and legs must be cut to shape before they can be attached to the chair frames.

Both the posts and the legs have narrow tapers on their edges. Since the tapers are only ¼" deep, use a belt sander to remove the material. Start by cutting the posts (D) and legs (E) to length. To cut the tapers in the posts and legs, you must measure and mark the guidelines carefully. Draw lines across the outside face of each leg and post, 5" up from their bottoms. Draw similar lines 6" down from the top of the posts. On the bottoms of the legs and posts (and the top of the posts), mark lines ¼" in from one edge, then connect these end lines to the face lines to form the shape of a narrow wedge. Set a compass to a 3" radius, and draw an arc on the rear edges of the posts. The arc should start at the top, tapered corner and extend 1" down the rear edge. Mount a belt sander sideways on your worksurface, and sand down the tapers and arcs on the legs and posts to

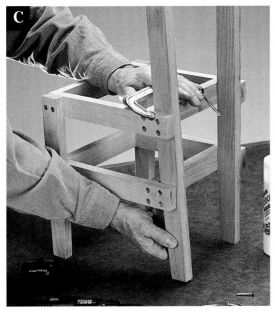

Clamp the posts in position on the back of the frame, and attach them.

Use the finished seat as a tracing pattern before cutting the slats to shape.

finished shape **(photo B).** To make sure the chairs don't wobble, do not remove any excess material from the bottom ends of the legs or posts.

ATTACH THE LEGS & POSTS. The legs are positioned at the front of the chairs, flush with the frames. The posts are located at the backs of the chairs, fitted between the sides and butted against the cross rails. When attaching the posts and legs, make sure the tapered edges are facing inward (see *Diagram*, page 31). Before you attach the legs and posts, clamp them in place to make sure they do not wobble. Apply glue to the legs, and position them against the frame sides so their front and top edges are flush with the front and top of the frames. Clamp the legs to the frames—use a combination square to check to make sure the legs are positioned on the frames at right angles. Drive evenly spaced counterbored wood screws through the legs and into the frames. Use glue

and wood screws to fasten the second frame between the legs so the top edge of the frame is 7" up from the bottoms of the legs. Clamp the posts to the sides **(photo C).** The top edge of the upper frame should be 12" above the bottoms of the posts. The lower frame should be 7" above the bottoms of the posts. Make sure the posts are butting flat against the cross rails, and fasten the posts to the sides with glue and counterbored wood screws. Drive the wood screws through the sides and into the posts. Use a sander to smooth over the edges of the posts and legs.

MAKE THE SLATS & SEAT. The seat is made from a 1 × 10 piece of oak. When you cut the seat to shape, the seat will have a 12"-long front edge and a 10"-long rear edge (see *Diagram*). Do not attach the seat immediately to the frame when you cut it to shape—you can use the finished seat shape as a tracing pattern for the slats. Cut the seat (G) to size. To cut the

seats to shape, mark points on one edge, 1" in from each side. Use a straightedge to draw diagonal lines from the opposing corners to these edge points. Cut along these diagonal lines with a jig saw, forming a seat with a 12"-long front edge and a 10"-long rear edge. Use a compass to draw ¾" radius semicircles on each seat corner. Cut along the curves with a jig saw to complete the seat. Use the finished seat as a pattern to cut the slats (F) to rough size. Position the top slat under the front of the seat so the edges are flush. Position the lower slat under the back of the seat with the edges flush. Trace around the seat with a pencil **(photo D),** and cut

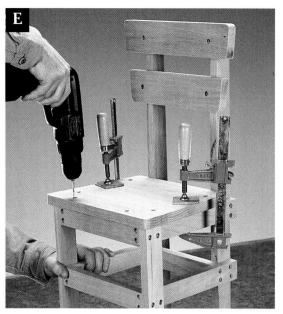

Extend the pilot holes through the seat into the frame to prevent splitting.

Smooth out the sharp edges on the apron with a power sander.

along the traced lines with a jig saw to complete the slats. Sand the edges with medium-grit sandpaper.

ATTACH THE SLATS & SEAT. The slats and seat are centered on the posts and frame, respectively, and attached with glue and wood screws. Before you attach the parts, however, drill pilot holes for the screws. Once you have the slats and seat clamped in position, extend the pilot holes through the slats and seat into the frame and posts to prevent splitting. Draw a center line down the slats, and drill two evenly spaced, counterbored pilot holes through each slat. The pilot holes should be centered 3⅝" from each side of their centers. Clamp the top slats to the posts so the top edges are flush. The top slat should extend 2" past the posts on each side. Fasten the slat with glue and #6 × 1¼" wood screws. Clamp the lower slat to the posts, 21¼" from the post bottoms, and fasten it to the posts with glue and wood

screws. Once the slats are attached, center the chair frame upside down on the seat top. Lightly trace the frame outline on the seat. Using the outline as a guide, drill evenly spaced, counterbored pilot holes through the seat for #6 × 1¼" wood screws. Center the seat over the frame, and clamp it in place. After extending the pilot holes into the frame **(photo E),** attach the seat with glue and wood screws. Glue and insert oak plugs into all the screw holes. When the glue has dried, sand the surfaces smooth, then finish-sand the entire chair with 180-grit sandpaper.

MAKE THE TABLE APRON. The table apron is a rectangular frame that anchors the legs and tabletop. Cut the apron sides (H) and apron ends (I) to size. Mark two evenly spaced centers for counterbored pilot holes at each end of the sides, ⅜" from each end. Drill pilot holes at each center for #6 × 1¼" wood screws. Fasten the apron ends between the

sides with glue and wood screws, driven through the pilot holes and into the apron ends. Smooth out the corners and edges with a sander to remove any rough spots or splinters **(photo F).**

MAKE THE LEGS. Each corner of the apron has an outside leg and inside leg. These legs are attached to each other in a butt joint. Both legs are tapered on one long edge. Start by cutting the outside legs (J) and inside legs (K) to length. Mark a point on one long edge of the outside and inside ends, 16" from the bottom—mark a point on the bottom, ¾" from the same long edge. Draw a diagonal line connecting both marks, and cut along the line with a jig saw to create a taper on each leg **(photo G).** Drill four counterbored pilot holes on the outside legs, ⅜" from the untapered edge of the outside legs. The pilot holes should be 6½" apart, starting 1" from the bottoms of the outside legs. Attach the outside legs to the in-

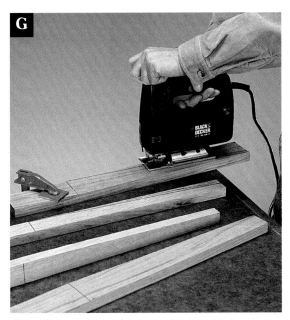

Use a jig saw to cut the tapers in the table legs.

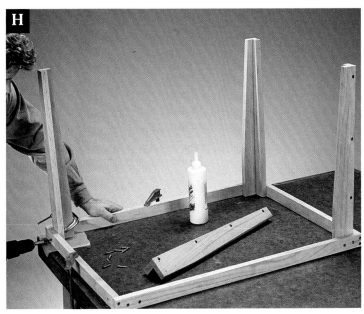

Attach a leg in each corner of the apron frame with glue and screws.

Attach the end nosing and side nosing with glue and finish nails.

side legs with glue and wood screws. Make sure that the untapered edges are flush. Apply glue to the legs and attach them to the table apron by driving counterbored wood screws through the apron and into each leg. Keep the tops of the legs flush with the top edges of the apron so the legs are even on their bottoms **(photo H).** When all the legs are attached, fill the screw holes with glued oak plugs and sand the surfaces to smooth out any rough spots.

MAKE THE TABLETOP. The tabletop is made from a solid piece of ¾"-thick oak plywood and is framed by the end nosing and side nosing. The nosing pieces are cut to fit around the tabletop from ¾" oak corner molding. Cut the tabletop (L), end nosing (M) and side nosing (N) to size. Use a miter box to miter-cut the ends of the nosing to fit around the tabletop. Center the tabletop over the apron, and attach it with evenly spaced, counterbored wood screws, driven through the tabletop into the top edges of the apron. Fill the counterbored screw holes with glued oak plugs, and sand the tabletop with medium-grit sandpaper. Fasten the end nosing and side nosing to the tabletop with glue and 1" wire brads **(photo I).** Drill pilot holes before driving the nails to avoid splitting the nosing. Set the nails and fill the holes with wood putty.

APPLY FINISHING TOUCHES. Finish-sand the table and chairs with fine (150- or 180-grit) sandpaper. Finish the project with two or three coats of satin-finish water-based polyurethane to protect the wood.

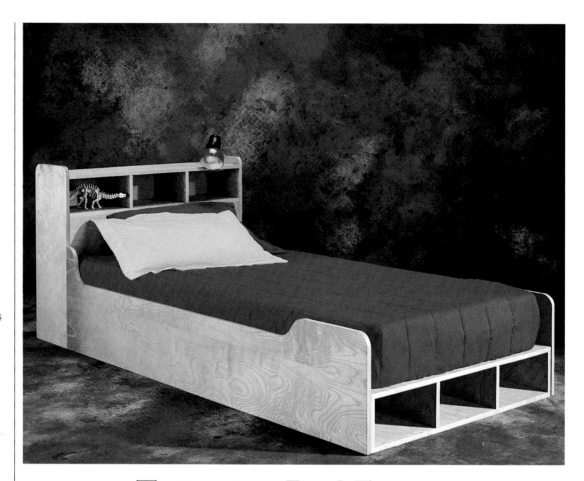

Twin-size Bed Frame

Soft, contemporary lines, warm wood tones, and plenty of built-in storage make this bed frame the centerpiece of any bedroom.

Beds, roller skates and winter coats have one thing in common: your child can outgrow them very quickly. While there is little you can do to make skates and coats last more than one or two seasons, a bed frame is a different story. This attractive bed

frame is designed with fun lines and six separate storage compartments, making it an exciting choice for a child's first real bed. But unlike many kids' beds that take fanciful design to an extreme, this bed can serve your child well into his or her teens, without seeming like an overgrown crib. And because it is made from two sheets of plywood, you can build it for a fraction of the cost of a less versatile, cheaply constructed bed that is purchased at a kids' furniture store.

This bed frame is sized to support a standard twin-size box spring and mattress. Be-

cause the box spring rests on wood cleats on the sides of the wood panel bedrails, you don't need to purchase any expensive bed rails or other bed hardware to make it. The main panels of the bed frame are connected with metal brackets, so the frame can be disassembled easily for transportation.

An assortment of storage cubbies and handy surfaces are well positioned for a reading lamp and alarm clock, as well as clothing that will not fit into closets or dressers. We chose a natural wood finish for our bed frame, but you can paint yours if you prefer.

CONSTRUCTION MATERIALS

Quantity	Lumber
1	¼" × 4 × 4' AB plywood
1	¾" × 4 × 8' birch plywood
3	2 × 2" × 8' pine

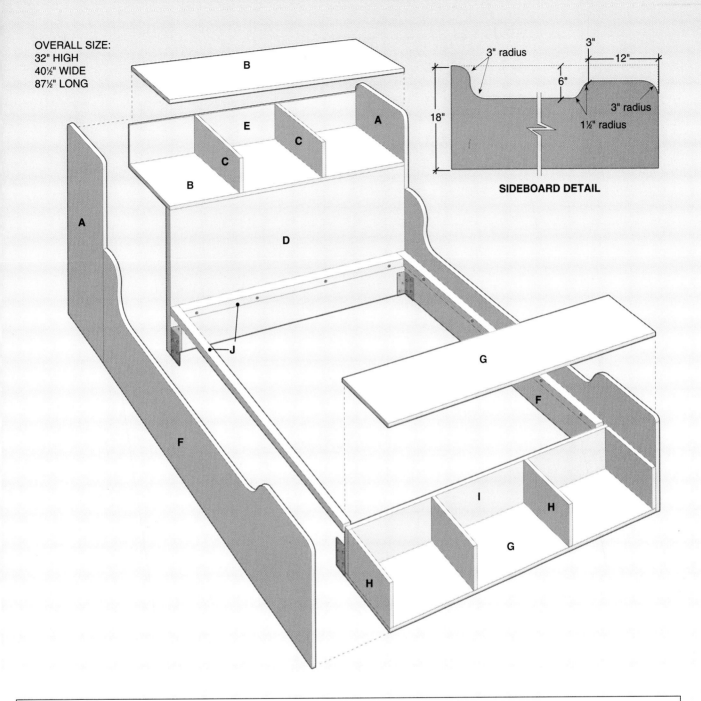

OVERALL SIZE:
32" HIGH
40½" WIDE
87½" LONG

SIDEBOARD DETAIL

3" radius
3"
12"
6"
18"
3" radius
1½" radius

Key	Part	Dimension	Pcs.	Material
A	Side panel	¾ × 11½ × 32"	2	Plywood
B	Headboard shelf	¾ × 11¼ × 39"	2	Plywood
C	Headboard divider	¾ × 11¼ × 6"	2	Plywood
D	Front panel	¾ × 39 × 23"	1	Plywood
E	Headboard back	¼ × 39 × 7¼"	1	Plywood

Cutting List

Key	Part	Dimension	Pcs.	Material
F	Sideboard	¾ × 18 × 76"	2	Plywood
G	Footboard shelf	¾ × 12 × 39"	2	Plywood
H	Footboard divider	¾ × 12 × 6"	4	Plywood
I	Footboard back	¾ × 39 × 7½"	1	Plywood
J	Ledger	1½ × 1½ × *	4	Pine

Materials: Wood glue, wood screws (#6 × ¾", #6 × 1½", #6 × 2"), 2d finish nails, metal corner brackets, birch veneer edge tape (50'), ⅜"-dia. birch wood plugs, finishing materials.

Note: Measurements reflect the actual size of dimensional lumber.
* Cut to fit

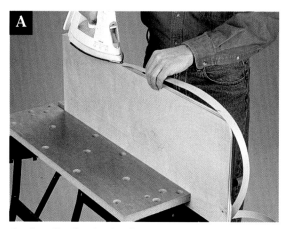

Apply self-adhesive birch veneer edge tape to plywood edges, using a household iron.

Draw a layout line for the bottom shelf across the side panels, even with the top of the front panel.

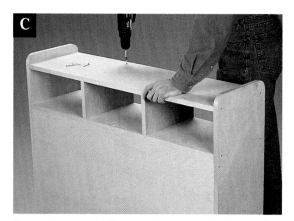

Secure the top shelf to the dividers and side panels with glue and counterbored screws.

Directions: Twin-size Bed Frame

MAKE THE SIDE PANELS. The side panels form the sides of the headboard unit. We used birch plywood for all plywood parts, but you could substitute oak plywood, or use a good cabinet grade (AB) fir plywood if you plan to paint the bed frame. Cut the side panels (A) to size from ¾"-thick birch plywood, then draw 3"-radius curves at the top corners of the side panels. Cut the curves with a jig saw. Lay one side panel on top of the other panel and trace the curved corners onto the other side panel. Cut the second set of curves. Smooth out the curves with a power sander and medium-grit sandpaper. Clean the edges of the panels thoroughly, then apply self-adhesive wood veneer edge tape to the edges of both panels. Cut the strips to length, then press them into place with a household iron **(photo A)**—the heat from the iron activates the adhesive. After the tape is set, trim off any veneer overhang with a utility knife, then sand the edges smooth.

MAKE & INSTALL THE FRONT PANEL & SHELVES. The front panel and shelves fit between the side panels to form the headboard. Cut the front panel (D) to size from ¾"-thick birch plywood. Fasten the front panel to the side panels so the bottoms of all panels are flush and the front edges of the side panels are flush with the front face of the front panel. Use glue and #6 × 1½" wood screws. Drive the screws through pilot holes that are counterbored to accept birch wood plugs (usually ⅜" dia.). Cut the headboard shelf (B) and the headboard dividers (C) to size, and smooth the cut edges with a sander. Apply edge tape to the front edges of the parts. Stand the headboard assembly upright and use a square to draw a reference line across the inside faces of the side panels, even with the top of the front panel **(photo B).** Position the lower headboard shelf on the top edge of the front panel, flush with the reference lines. Drill counterbored pilot holes through the side panels into the ends of the shelf, then fasten the shelf to the side panels with glue and #6 × 1½" wood screws. Measure 13" from each end of the bottom shelf and draw lines across the shelf. Position a headboard divider inside each mark, and secure the dividers with glue and screws driven up through the bottom shelf. Position the upper headboard shelf on the dividers, with the ends of the shelf flush against the inside faces of the side panels. Make sure the dividers are exactly perpendicular to the lower headboard shelf, then fasten them to the upper shelf with glue and counterbored screws **(photo C).** Also drive screws through the side panels and into the ends of the upper headboard shelf. Cut the headboard back (E) to size from ¼"-thick plywood, and fasten it to the back of the headboard with 2d finish nails **(photo D).**

Fasten the ¼"-thick plywood back panel to the headboard using 2d finish nails.

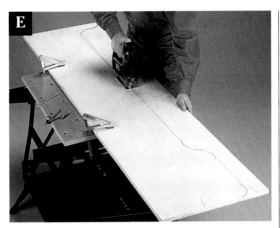

Draw contoured cutting lines on the top edges of the sideboards, then cut with a jig saw.

MAKE THE FOOTBOARD. The footboard is similar to the headboard. Cut the footboard shelves (G) and footboard dividers (H) to size. Apply wood veneer edge tape to the front edges. Draw a reference line 13" from each end of the bottom shelf. Position a footboard divider inside each reference line, and at the ends of the footboard shelves. Secure the dividers with glue and screws driven through the shelves and into the tops and bottoms of the dividers. Make sure the front edges of the parts are flush. Cut the footboard back (I) to size and fasten it to the backs of the shelves and dividers with glue and screws.

BUILD THE SIDEBOARDS. The sideboards connect the headboard to the footboard, while supporting ledgers that hold the box spring. Cut the sideboards (F) to size. Lay out the curves and contours along the top edge (see *Diagram,* page 37) of one sideboard. Cut the top edge of a sideboard to shape with a jig saw **(photo E).** Smooth the edges with a sander, then use the sideboard as a template for tracing a matching contour onto the other sideboard. Cut and sand

the second sideboard, then apply wood veneer edge tape to the top and front edges. Fit the headboard and footboard between the sideboards, so the outside of the footboard is flush with the front ends of the sideboards, and the headboard unit is butted against the back edges of the sideboards. Use duct tape to fasten the parts together temporarily, then check to make sure everything is square. Measure from the front of the headboard to the back of the footboard, and cut two 2 × 2 pine ledgers (J) to this length. Remove the sideboards, and fasten the ledgers to the inside faces so they are 7½" up from the bottoms of the sideboards, flush with the back edges. Use #6 × 1½" screws fastened at 8" intervals to attach the ledgers. Replace the sideboards, and measure between the ledgers on the sideboards to find the right length for the ledgers that are attached to the headboard and footboard. Cut and attach these ledgers, so the tops of all ledgers are flush.

ASSEMBLE THE BED. Attach the headboard and footboard to the sideboards with metal corner-brackets at each joint **(photo F).** Set your twin-size

box spring onto the ledgers to make sure everything fits.

APPLY THE FINISH. Disassemble the frame, then fill all counterbores with wood plugs. Sand the plugs flush with the surfaces, then finish-sand all wood surfaces with 150- to 180-grit sandpaper. Apply stain or paint (we used a light oak stain). Apply finish materials to all wood surfaces. Apply at least two coats of topcoat (we used water-based polyurethane with a satin gloss). Reassemble the bed frame in the room where it will be used.

Fasten the sideboards to the headboard and footboard with corner brackets and screws.

Backboard Hamper

With this simple project, your child will be doing laundry lay-ups and slam-dunking dirty denims in no time.

PROJECT
POWER TOOLS

This playful project gives youngsters an added incentive to keep their room tidy by making household chores fun.

CONSTRUCTION MATERIALS

Quantity	Lumber
1	2 × 4" × 4' pine
2	1 × 4" × 8' pine
1	1 × 2" × 8' pine
1	¾" × 2 × 4' plywood

The first basketball game was played with a peach basket on a post for a goal. We expanded on that piece of sports trivia by designing this backboard hamper with a laundry basket as a goal. By tapping into a child's instinct to play, you can make a game out of putting dirty clothes in their proper place (you may know a few adults who could put this project to good use as well).

The parts used to make this backboard hamper are made from pine and plywood. Because it is such a prominent feature, use extra care when cutting the semicircular backboard. The shelf that supports the laundry basket can be raised or lowered during construction of the project to accommodate a basket with different dimensions than the one we used, which is 16 × 16 square, and 10½" tall.

OVERALL SIZE:
48" HIGH
19" DEEP
20" WIDE

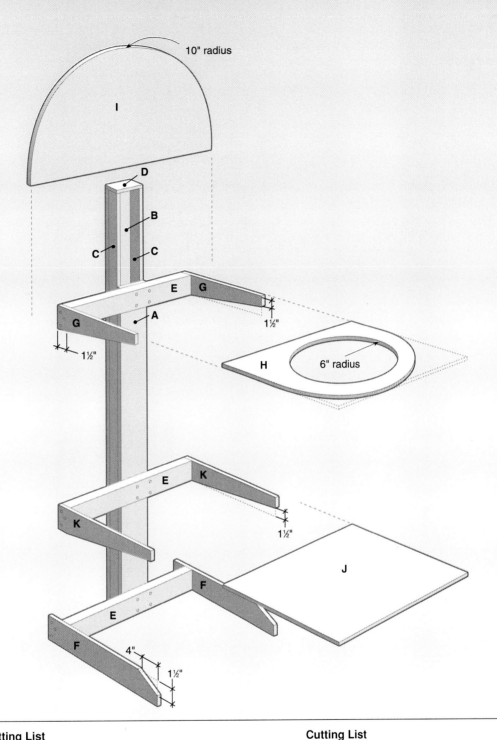

10" radius

6" radius

1½"

1½"

1½"

4"

1½"

	Cutting List			
Key	**Part**	**Dimension**	**Pcs.**	**Material**
A	Post front	¾ × 3½ × 36"	1	Pine
B	Post back	¾ × 3½ × 46"	1	Pine
C	Post side	¾ × 1½ × 46"	2	Pine
D	Post cap	¾ × 3½ × 2¼"	1	Pine
E	Crossbar	1½ × 3½ × 14½"	3	Pine
F	Foot	¾ × 3½ × 19"	2	Pine

	Cutting List			
Key	**Part**	**Dimension**	**Pcs.**	**Material**
G	Rim support	¾ × 3½ × 8"	2	Pine
H	Rim	¾ × 16 × 16"	1	Plywood
I	Backboard	¾ × 12 × 20"	1	Plywood
J	Basket shelf	¾ × 16 × 16"	1	Plywood
K	Shelf support	¾ × 3½ × 16"	2	Pine

Materials: Glue, wood screws (#6 × 1⅝", #6 × 2"), finishing materials, 16 × 16 × 10½" laundry basket.

Specialty Items: Combination square.

Note: Measurements reflect the actual size of dimensional lumber.

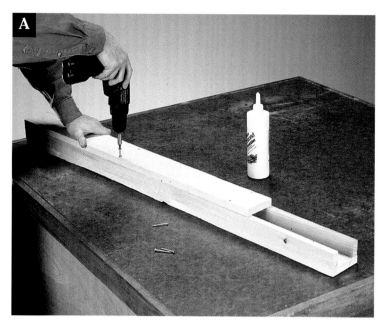

Make the central post by fastening the post sides between the post front and the post back, using glue and screws.

Attach the cap to the top of the post with glue and wood screws.

Directions:
Backboard Hamper

BUILD THE POST. The post is attached to the base to support the shelf, rim and backboard. It is made from four pieces of pine butted together to form a box. Cut the post front (A), post back (B) and post sides (C) to size. Sand all parts with medium-grit sandpaper to smooth out any rough edges. Keeping the bottom and side edges flush, fasten the sides between the front post and back post with glue and #6 × 1⅝" wood screws, driven through the posts and into the sides **(photo A).** Drill pilot holes for all screws, and countersink them so you can cover the screw heads with wood putty.

ATTACH THE POST & BASE. The base for the backboard hamper consists of two feet attached to a crossbar on the central post. Cut a crossbar (E) and both feet (F). Mark a point on each foot, 1½" down from

the top edge, on the front end. Mark another point 4" in from the front, top corner on the top edge. Draw a line connecting the points on each foot, and cut along the lines with a circular saw or jig saw to cut the feet to shape. Sand the parts smooth. Attach a crossbar to the post front, flush with the bottom edge, making sure the crossbar overhangs the same amount on each side of the post. Use four #6 × 2" wood screws, driven through counterbored pilot holes, to attach the crossbar (see *Diagram*, page 41, for screw placement). Attach one foot to each end of the crossbar, using glue and #6 × 2" screws. Each foot should extend 3" past the back side of the crossbar.

MAKE THE BACKBOARD. Cut a piece of ¾"-thick plywood to 12 × 20" to make the backboard (I). Mark the centerpoint of the backboard on the top edge, then measure straight down 10" and mark a point. Tack a finish

nail at the point, then tie a string to the nail. Tie a pencil to the other end of the string, so the tip of the pencil is even with the centerpoint on the top edge. Draw a semicircular curve on the plywood—the top of the curve should be flush with the top of the plywood, and the ends of the semicircle should be level with the nail, 2" up from the bottom edge. Cut along the cutting lines with a jig saw, then sand the edges of the cut smooth with a sander and medium-grit sandpaper.

INSTALL RIM SUPPORTS. Cut another crossbar (E) and both rim supports (G). On the supports, mark a point on one end of each support, 1½" down from the top edge. Mark another point on the bottom edge, 1½" in from the back end of the support. Draw a cutting line connecting the points on each support, and cut along the cutting lines with a circular saw. Sand the supports smooth. Attach the rim supports to the ends of the crossbar with glue

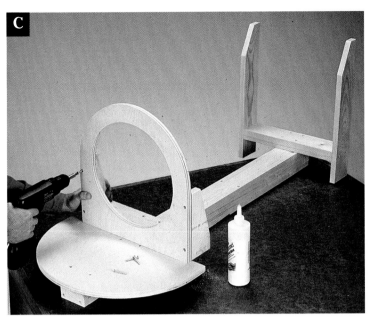

C

Center the hoop on the arms, making sure its rear edge is against the backboard, and fasten it with glue and wood screws.

and #6 × 2" wood screws. With the base and post upright, center the crossbar on the front of the post with the overhang equal at each side of the post. The top of the crossbar should be flush with the top of the post front (which is 10" down from the top of the post back). Attach the crossbars the same way you attached the crossbar for the feet. Cut the post cap (D), and attach it to the top of the post with glue and wood screws **(photo B).**

MAKE THE RIM. Cut the rim (H) to size from ¾"-thick plywood. Mark the centerpoint on the back edge, then measure straight down 8" and mark another point. Drive a nail at the point, and use the string compass technique from the previous section to draw a pair of semicircles with the nail as a centerpoint: draw a full circle with a 6" radius to make the opening in the rim; and draw a semicircle with an 8" radius to

mark the cutting line for the rounded front of the rim. Drill a starter hole in the center circle, and cut with a jig saw. Cut along the outer cutting lines to form the rim front. Sand all edges smooth.

ATTACH THE BACKBOARD & RIM. Set the backboard in position, with its straight edge resting on top of the post front board. Make sure the backboard is centered on the post. Attach it to the front edges of the post sides with glue and #6 × 2" wood screws driven through counterbored pilot holes. Position the rim on the rim supports so the straight edge is butted against the backboard. Make sure the rim is centered from side to side on the backboard, then attach it with glue and #6 × 2" wood screws driven through counterbored pilot holes **(photo C).**

INSTALL THE BASKET SHELF. On the backboard hamper shown here, we installed a shelf below the rim to hold a

10½"-high laundry basket. If you wish to use a larger laundry basket, adjust the height of the shelf. Or, if you'd rather use a laundry bag than a laundry basket, you can eliminate the shelf. Cut the basket shelf (J) to size and sand the edges smooth. Cut the third crossbar (E) and the shelf supports (K). Draw cutting lines on the supports , starting 1½" up from the bottom at the front end of each support, and running in a straight line to a point 1½" up from the bottom of the lower back corner. Cut with a circular saw. Attach the supports to the ends of the crossbar, and attach the crossbar to the post front (we set ours 20½" up from the floor). Attach the basket shelf to the shelf supports.

APPLY FINISHING TOUCHES. Fill all counterbores and exposed plywood edges with putty, then sand the surfaces smooth. Wipe the surfaces clean, then paint the entire project (we used blue enamel paint on everything but the backboard, which we painted white). We also masked off and painted a 12 × 12" square target onto the backboard **(photo D).**

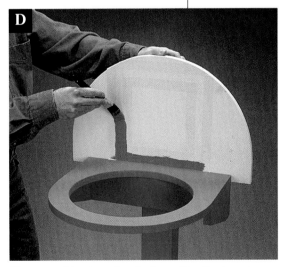

D

Use a masking tape border to outline the target area when painting the backboard.

Game Table

*A cutout game board and concealed storage box make this child-size
game table a great way to keep your parent-size tables clear.*

CONSTRUCTION MATERIALS

Quantity	Lumber
1	¾" × 4 × 8' plywood

The perfect table for any board game, this table features ample storage, in addition to sturdy construction and specialized design. The top contains a cutout game table that easily pulls out to reveal a storage space below. This concealed storage area is perfect for holding the small game pieces that always seem to end up on the floor or under the couch. We designed the game table at a convenient height for most children. The kids will appreciate this game table because it is fun. You will appreciate it because it gives the children a specific area just for games, keeping the coffee table and the living room floor free from clutter.

OVERALL SIZE:
20" HIGH
36" WIDE
35" LONG

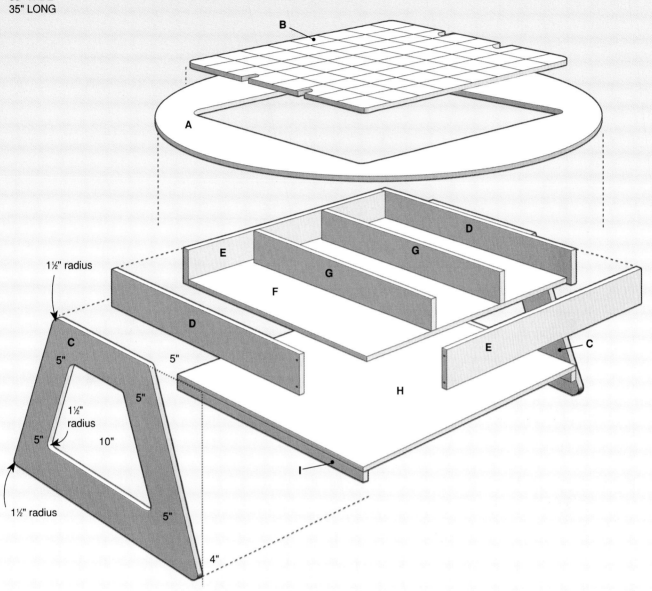

Cutting List				
Key	**Part**	**Dimension**	**Pcs.**	**Material**
A	Top	¾ × 36 × 36"	1	Plywood
B	Game board	¾ × 22¼ × 22¼"	1	Plywood
C	Leg	¾ × 18 × 24"	2	Plywood
D	Side	¾ × 4 × 24"	2	Plywood
E	End	¾ × 4 × 22½"	2	Plywood

Cutting List				
Key	**Part**	**Dimension**	**Pcs.**	**Material**
F	Bottom	¾ × 22½ × 22½"	1	Plywood
G	Divider	¾ × 3¼ × 22½"	2	Plywood
H	Fixed shelf	¾ × 22½ × 24"	1	Plywood
I	Cleat	¾ × 1½ × 20"	2	Plywood

Materials: Glue, wood screws (#6 × 1¼", #6 × 2"), finishing materials.

Note: Measurements reflect the actual size of dimensional lumber.

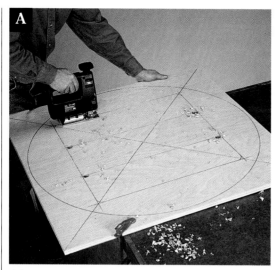

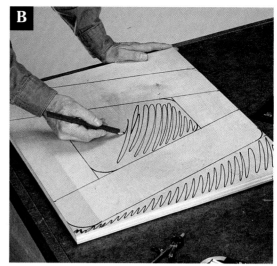

After drawing the guidelines, cut the game board cutout from the top with a jig saw.

Shade the waste areas on the plywood to avoid confusion when cutting the leg to shape.

Directions: Game Table

BUILD THE TOP. The top starts as a large square. You must lay out cutting guidelines carefully to achieve the finished circular shape and built-in game board. Begin by cutting the top (A) to size from ¾"-thick plywood. Mark the centers of each edge on the 36 × 36" square top, and use a straightedge to draw lines connecting the centers on opposing edges. The point where the lines intersect is the center of the board. Draw marks along each edge, 11⅛" to each side of the edge centers. Use a straightedge to draw lines connecting these marks on opposing edges, creating a square with 22¼" sides on the top. This square is the outline of the game board (B). To cut the top to its finished circular shape, you need to build a makeshift compass. Drive a wire nail into the table centerpoint. Tie one end

of a piece of string around the nail and the other around a pencil—the string should measure 17½" in length when tied to the nail and pencil. Pull the pencil outward until the string is taut, and mark the line on the material as you circle the nail. The result is a circle with a 17½" radius. Use a compass to draw a curve with a ¾" radius in each corner of the game board cutting lines. To make finger grips for the game board, mark two points along two opposing sides of the square, 7" in from the corners. Drill 1"-dia. holes at each of the points, then cut along the drawn square with a jig saw to make the game board **(photo A)**. Cut around the circle lines to finish the top shape. Lightly sand all

Insert the dividers and attach them between the ends with glue and wood screws.

the edges. Do not sand the edges of the cutout extensively—too much sanding will create large gaps between the game board and the top.

MAKE THE BASE. The information on cutting the legs (C) to shape can be found in the *Diagram*, page 45. When you have one leg cut to shape, just trace around it to create the second leg. Lay out the guidelines for cutting one leg to shape (see *Diagram*, page 45). Use a compass to draw 1½"-radius curves on all the corners.

TIP

If you'd rather not cut the round top to shape, you can purchase 36"-dia. round particleboard tabletops at most building centers.

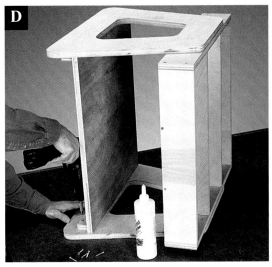

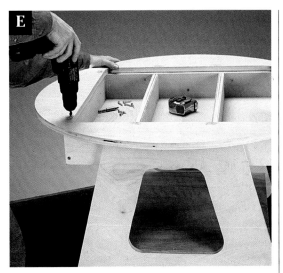

Drive wood screws through the cleats and into the legs to fasten them beneath the fixed shelf.

Fasten the top to the storage box with glue and countersunk wood screws.

To avoid confusion when cutting, shade the waste areas **(photo B).** Cut the leg to shape with a jig saw. Draw the cutting guidelines for the second leg, using the finished leg as a tracing template.

MAKE THE STORAGE BOX. The storage box is a simple frame with a plywood bottom that fits directly under the top. Start by cutting the sides (D), ends (E) and bottom (F) to size. Fasten the ends between the sides with glue and countersunk #6 × 2" wood screws, driven through the sides and into the ends. Make sure the outside faces of the ends are flush with the side edges. Position the bottom inside the frame formed by the ends and sides, and fasten it with glue and screws, flush with the bottom of the frame. Cut the dividers (G) to size. To attach the dividers, mark lines on the ends, 6½" from the sides. Drill evenly spaced pilot holes for #6 × 2" wood screws on this line. Fasten the dividers between the ends with glue and screws **(photo C).**

ATTACH THE LEGS. Draw lines on the top, side edges of the storage box, 5" from each end. Align one leg between these lines so the top edges are flush. Apply glue to the parts, and fasten the leg to the storage box with #6 × 1¼" wood screws, driven through the box side and into the leg. Repeat this procedure with the other leg. Cut the fixed shelf (H) and cleats (I) to size. Use glue and countersunk wood screws to attach the fixed shelf between the legs. The top of the fixed shelf should be 4" from the bottom of the leg. Position a cleat directly under each side of the fixed shelf. With the cleats butting against the bottom of the fixed shelf, fasten them to the inside of the legs with glue and countersunk screws **(photo D).**

APPLY FINISHING TOUCHES. Center the top on the storage box, and drive countersunk #6 × 2" wood screws through the top and down into the top edges of the storage box **(photo E).** Fill all the countersunk screw holes with wood

putty, and finish-sand all the surfaces with fine-grit sandpaper. Place the game board in the top cutout. Make sure it is well supported and easy to lift out. Trim the edges and deepen the finger holes by sanding with a drum sander, if necessary. Remove the game board. Prime the surfaces, then paint the game table (we used a semigloss latex enamel paint). Although you can leave the game table blank, we sponge-painted it with rows of 1½"-wide squares for playing checkers or chess **(photo F).**

Sponge-paint a checkerboard onto the game board for built-in fun.

PROJECT
POWER TOOLS

Chest of Drawers

*Why pay high store prices when you can build this functional,
inexpensive chest of drawers yourself?*

CONSTRUCTION MATERIALS

Quantity	Lumber
1	¾" × 4 × 8' plywood
1	¾" × 2 × 4' plywood
1	½" × 4 × 4' plywood
1	¼" × 2 × 4' plywood
1	1 × 2" × 4' pine
1	¾ × ¾" × 8' cove molding
1	½ × ½" × 8' quarter-round

Children's furniture is notoriously expensive, but you can save money by building this simple set of drawers for a fraction of the cost of a store-bought dresser. To make it you'll use only the most basic carpentry techniques to achieve a result with a clean, attractive appearance. You won't find a chest of draw- ers that's cheaper or easier to make anywhere. This chest of drawers features all wood parts—there is no need to pur- chase metal drawer slides or other expensive types of hard- ware. Not quite full-size, this chest of drawers will fit nicely into your child's bedroom, while still offering plenty of clothing storage space.

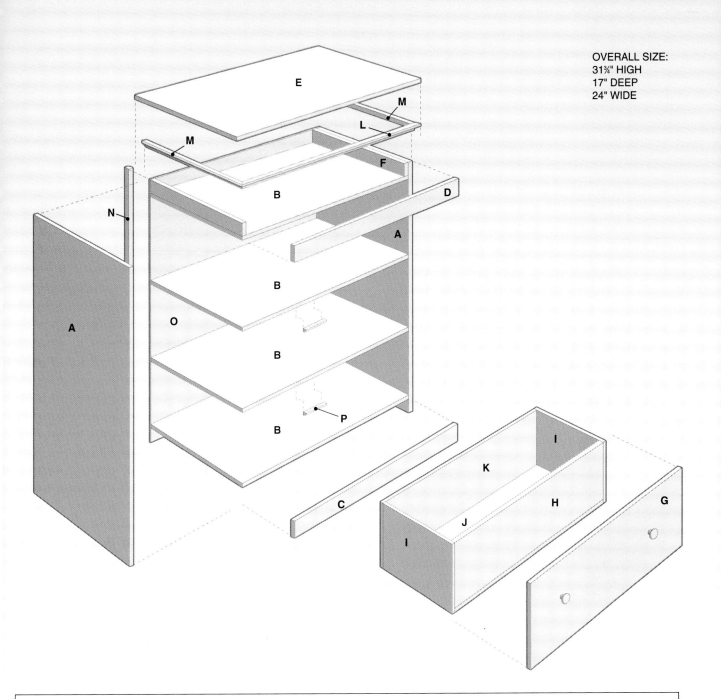

OVERALL SIZE:
31¾" HIGH
17" DEEP
24" WIDE

	Cutting List						Cutting List			
Key	**Part**	**Dimension**	**Pcs.**	**Material**		**Key**	**Part**	**Dimension**	**Pcs.**	**Material**
A	Side	¾ × 15¼ × 31"	2	Plywood		**I**	Drawer side	½ × 8 × 14¼"	6	Plywood
B	Duster	¾ × 14½ × 20½"	4	Plywood		**J**	Drawer bottom	¾ × 13¾ × 19¼"	3	Plywood
C	Bottom rail	¾ × 2 × 22"	1	Plywood		**K**	Drawer back	½ × 8 × 19¼"	3	Plywood
D	Top rail	¾ × 1½ × 22"	1	Plywood		**L**	Front cove	¾ × ¾ × 23½"	1	Molding
E	Top	¾ × 17 × 24"	1	Plywood		**M**	Side cove	¾ × ¾ × 16¾"	2	Molding
F	Top cleat	¾ × 1½ × 14½"	2	Pine		**N**	Back cleat	½ × ½ × 29⅜"	2	Molding
G	Drawer face	¾ × 9 × 22"	3	Plywood		**O**	Back	¼ × 20½ × 29⅜"	1	Plywood
H	Drawer front	¾ × 7¼ × 19¼"	3	Plywood		**P**	Stop block	½ × ½ × 4"	3	Molding

Materials: Glue, wood screws (#6 × 1¼", #6 × 2", #4 × 1"), 4d finish nails, 1" wire nails, plastic drawer glides (24), 2"-dia. cabinet drawer knobs (6), finishing materials.

Note: Measurements reflect the actual size of dimensional lumber.

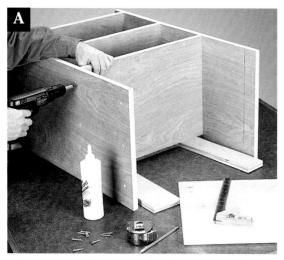

The duster panels support the drawers. Attach them with glue and screws driven through the sides.

Attach cleats made from ½" quarter-round molding at the outside edges of the back panel to give it extra strength.

Tack plastic drawer glides to the inside faces of the drawer openings to eliminate friction.

Directions: Chest of Drawers

BUILD THE CHEST. A chest of drawers is exactly that—a wooden chest that contains several storage drawers. The chest is a simple plywood box with shelves, called "dusters," that support the drawers and give the chest lateral strength. Cut the sides (A) and dusters (B) to size from ¾"-thick plywood (we used AB plywood). Cut the back panel (O) from ¼"-thick sheet goods (we used ¼"-thick plywood, but you can use inexpensive hardboard).

Sand all parts smooth. Mark guidelines for the duster positions on the inside faces of the sides, 1⅞", 10¾", 19⅞" and 29" from the bottoms of the sides. The dusters are positioned with their bottom faces on these lines and are flush with the front edges of the sides. Stand the sides on their back edges, and fasten the dusters between the sides with glue and #6 × 2" wood screws **(photo A).** Set ¾"-thick spacers beneath the dusters so the front edges are flush wth the front edges of the sides. Drive the screws through pilot holes that are countersunk just deeply enough so you can fill them with wood putty to cover the screw heads. Lay the back in place against the back edges of the dusters, and fasten it with 1" wire nails, driven through the back and into the dusters. Cut the back cleats (N) to size from ½" quarter-round molding, and attach it to the back edges of the sides with 1" brads **(photo B).**

ATTACH THE RAILS & TOP. Cut the bottom rail (C) and top rail (D) to size from plywood. Position the top and bottom rails so the ends are flush with the outside faces of the sides. The top of the top rail should be flush with the tops of the side panels, and the bottom rail should be flush with the bottoms of the sides. Fasten the rails to the front edges of the sides and dusters with glue and wood screws. Cut the top (E) from plywood and cut the top cleats (F) from 1 × 2 pine. Attach the cleats to the inside faces of the side panels, flush with the top of the chest, by driving 1¼" screws through the cleats and into the sides. Position the top onto the chest so it is flush with the back, and the overhang is the same at each side. Attach the top with glue and #6 × 2" screws driven through counterbored pilot holes into the tops of the cleats. Use a router with a ¼" roundover bit, or use a power sander, to smooth out the sharp edges of the top. Tack four ⅛"-thick plastic drawer glides to the ends of each drawer opening to eliminate friction between the drawers and the chest **(photo C).**

Fasten the sides to the front with wire nails driven through pilot holes in the sides.

Drive screws through the knob holes in the drawer face to temporarily attach it to the drawer front.

BUILD THE DRAWER BOXES. The drawers are simple boxes with decorative front faces attached. Cut the drawer fronts (H), drawer bottoms (J), drawer sides (I) and drawer backs (K) from the appropriate thickness of plywood for each part. Fasten a drawer bottom to a drawer front with glue and counterbored #6 × 1¼" wood screws, driven through the bottom and into the bottom edge of the drawer front. Fasten the sides to the front and bottom with glue and 1" wire nails, driven through the sides **(photo D).** Fasten the drawer backs to the back edges of the drawer sides and bottom, using glue and 1" wire nails.

ATTACH THE DRAWER FACES & KNOBS. We attached two 2"-dia. wood drawer knobs to the face of each drawer. Most drawer knobs have a bolt in the center that is inserted through the drawer front and secured with a nut. Cut the drawer faces (G), and round over all the edges with a router or sander. Mark drilling centers for guide-holes, centered 4" from each end of each drawer face. Drill

guide holes the same size as the bolts through the centers. Tape a ¼"-thick spacer to the back of each drawer, and slide the drawers into the openings in the chest. The glides should keep the drawers centered on each duster, and the spacers will ensure that the front edges of the drawer fronts stay aligned with the front edges of the sides. Start with the bottom drawer when attaching the drawer faces. Maintain a ⅛"-wide space between the bottom of the drawer face and the bottom rail. Tape or clamp the drawer face in place on the drawer front. Drive 1¼"-long screws through the knob guide-holes and into the drawer front to hold the drawer face temporarily **(photo E).** Remove the drawer, and drive #6 × 1¼" wood screws through the back side of the drawer front and into the drawer face. Remove the screws in the knob holes, and position the bottom drawer in the frame. Attach the drawer faces to the two remaining drawers, maintaining ⅛"-wide gaps between the drawer faces. Attach knobs to the drawer faces.

APPLY FINISHING TOUCHES. Miter-cut the front cove (L) and side cove (M) with matching 45° angles, and install them in a frame where the top is joined to the chest, using 4d finish nails. Cut the stop blocks (P) to size. Remove the drawers, and fasten the blocks on the duster bottoms, ¾" from their front edges, using 1" brads **(photo F).** Fill all nail and screw holes and exposed plywood edges with wood putty, then sand all the surfaces smooth. Prime and paint all wood surfaces.

Turn the frame upside down, and use 4d finish nails to attach the stop blocks.

Kid-size Picnic Table

*This picnic table is loaded with sturdiness and charm,
but in a small package designed just for kids.*

A scaled-down version of an adult picnic table, this project is sure to appeal to all children. But with its stylish lines and warm, attractive appearance, you may want to build a scaled-up version for yourself.

But don't worry that your kids will outgrow this table too quickly. Although it is kid-size,

it is large enough to provide comfortable seating for kids of all ages.

Made completely from dimensional cedar lumber, this kid-size picnic table is lightweight despite its strength and durability. With a protective finish like linseed oil, it will maintain its warm cedar tones for many seasons.

OVERALL SIZE:
26" HIGH
48" WIDE
48" LONG

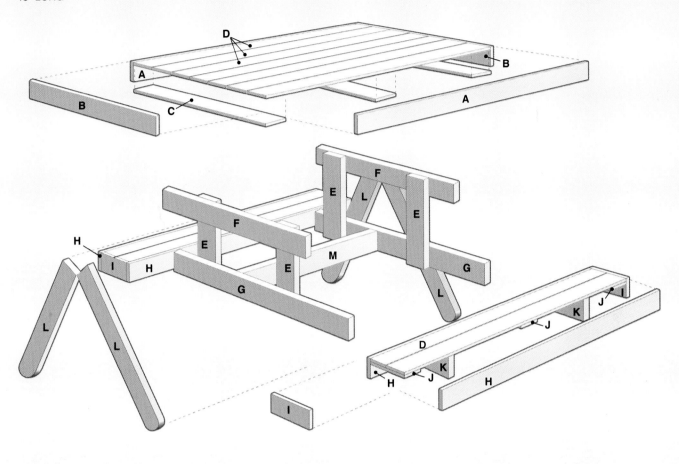

	Cutting List					Cutting List			
Key	Part	Dimension	Pcs.	Material	Key	Part	Dimension	Pcs.	Material
A	Table side	⅞ × 3½ × 48"	2	Cedar	H	Seat cap	⅞ × 3½ × 48"	4	Cedar
B	Table cap	⅞ × 3½ × 25¼"	2	Cedar	I	Seat end	⅞ × 3½ × 7¼"	4	Cedar
C	Table stringer	⅞ × 3½ × 25¼"	3	Cedar	J	Seat stringer	⅞ × 3½ × 7¼"	6	Cedar
D	Slat	⅞ × 3½ × 46¼"	11	Cedar	K	Seat cleat	⅞ × 3½ × 7¼"	4	Cedar
E	Temporary post	1½ × 3½ × 15¾"	4	Cedar	L	Leg	1½ × 3½ × 31"	4	Cedar
F	Top rail	1½ × 3½ × 25¼"	2	Cedar	M	Stretcher	1½ × 3½ × 26¼"	1	Cedar
G	Seat rail	1½ × 3½ × 46¼"	2	Cedar					

Materials: Moisture-resistant wood glue, galvanized deck screws (1¼", 2", 2½").

Specialty items: Combination square.

Note: Measurements reflect the actual size of dimensional lumber.

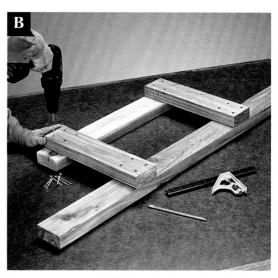

Slip a pair of slats under the table stringers as spacers, then drive deck screws to attach the side caps to the stringers.

Temporary posts hold the rails in position during assembly. Remove them when assembly is over.

Directions:
Kid-size Picnic Table

BUILD THE TABLETOP. The tabletop is formed by attaching seven cedar slats inside a 1 × 4 cedar frame. Cut all 11 1 × 4 cedar slats (D)—four will be used to make the seats—as well as the side caps (A), table caps (B) and table stringers (C). Sand all parts smooth. Set four slats aside. Position the table caps between a pair of side caps to form a rectangular frame. Fasten the rails between the side caps with moisture-resistant wood glue and 2½" deck screws, driven through the side caps and into the ends of the rails. Drill pilot holes before driving all screws, countersinking the pilot holes slightly so the screw heads will be recessed. With the frame lying flat on your worksurface, lay slats just inside each side cap to use as spacers while you attach the table stringers. Arrange the stringers inside the frame, so the two end stringers are flush against the inside faces of the table caps. The middle stringer should be centered

between the outer stringers. Drive deck screws through the side caps into the ends of the stringers **(photo A).** Turn the frame over on your worksurface, and arrange seven slats on top of the stringers, keeping a ⅛"-wide gap between slats (use 10d nails as spacers). Make sure the ends of the slats are butted against the inside faces of the table caps (if the slats are a little short, that's okay; just try to keep the ends even on each side). Clamp a board across each end of the tabletop to hold the slats in place, and turn the assembly facedown on your worksurface. Drive two 1¼" deck screws through each stringer and into each slat.

BUILD THE TABLE SUPPORTS. The table is supported by two cross rails that are attached temporarily to posts during assembly, then fastened permanently to the legs (the posts are then removed). A second pair of rails below these cross rails extend out past the tabletop to help support the seats. The lower rails are connected with a stretcher. Cut the posts (E),

top rails (F), seat rails (G) and stretcher (M) to size. Draw guidelines across the seat rails, 11¾" from each end. Use deck screws to fasten two posts to each seat rail, so their outside edges are on the guidelines, and their ends are flush with the outside edges of the rail. Fasten a top rail to the free ends of the posts, so the outside edge of each post is 1½" from the ends of the top rail **(photo B).** Turn the tabletop facedown on your worksurface, and position the top rails inside the tabletop frame, so the inside face of each top rail is 10" from the nearer table cap. Make sure the posts are inside the top rails. Apply glue, then drive two 1¼" deck screws through the side caps and into both ends of each top rail **(photo C).** Cut the stretcher (M) to size and position it between the two seat rails. Make sure the posts are perpendicular to the slats and the stretcher is centered end to end on the seat rails, then fasten it with glue and 2½" deck screws, driven through the seat rails into the ends of the stretcher.

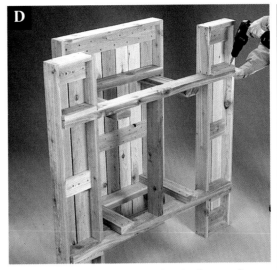

Make sure the inside face of each top rail is 10" from the tabletop ends, and fasten the parts.

The seat rails fit through notches in the seat frames and are attached to the seat cleats.

MAKE THE SEATS. The seats are built in much the same way as the tabletop. Cut the seat caps (H), seat ends (I), seat stringers (J) and seat cleats (K) to size. You'll also need the four slats (D) you set aside earlier. Arrange the seat caps in two pairs, and position a seat end between both ends of each pair, so their edges are flush. Attach the seat ends to the seat caps with glue and 2" deck screws, forming two seat frames. Arrange three seat stringers facedown within each frame so a stringer is flush with each seat end and the third stringer is centered between the ends. Use slats as spacers beneath the stringers (see photo A), and attach the stringers with glue and 2" deck screws driven through the seat caps and into the ends of the stringers. Slip two slats in each frame, flush against the seat caps, with a gap between slats. Fasten the slats with glue and 1¼" deck screws driven through the stringers and into the slats. Now, you'll need to cut notches into the inside faces of

the inner seat caps to fit around the seat rails. Mark two 1½"-wide × 2⅜"-high notches for the seat rails on the face of each inner seat cap, starting 9⅜" in from the seat ends. Attach a seat cleat between the seat caps at the outside edge of each notch (see *Diagram*, page 57). Use glue and 2½" deck screws driven through the side caps and into the ends of the seat cleats. This completes the preparation of the seats. Slip the notches in each seat over the seat rails, so the seat rails butt against the inside faces of the outer seat caps. Attach the seat rails to the seat cleats with glue and 2½" deck screws **(photo D).**

MAKE THE LEGS. Cut the legs (L), then use a compass to draw a curved cutting line with a 1¾" radius at one end of each leg. Cut along the cutting lines with a jig saw. Stand the table on end, and draw a line to mark the center of each top rail. Position the legs on the outside faces of the rails, so the top inside corner of each leg touches the centerline, and the top outside corner of each leg is pressed against the under-

side of the tabletop. Each leg should also fit against the outer face of a seat rail and press against the inside bottom edge of the seat cap. Fasten the legs in these positions with glue and 2½" deck screws **(photo E).** Remove the posts.

APPLY FINISHING TOUCHES. Sand all the wood edges and surfaces, then apply a non-toxic protective finish (we used linseed oil). Refresh the finish annually.

Each leg should fit against the underside of the tabletop and the lower inside edge of the seat cap.

Kiddie Craft Center

A work center and a storage unit are combined in this craft center, making hobby work easy and fun.

CONSTRUCTION MATERIALS

Quantity	Lumber
2	1 × 4" × 8' pine
1	2 × 4" × 8' pine
1	1 × 10" × 6' pine
1	1 × 2" × 10' pine
1	¾" × 4 × 4' plywood
1	¼" × 2 × 4' pegboard
1	1⅛ × 1⅛" × 8' corner molding

Whether your child's hobby is stamp collecting, model building, or anything in between, this craft center will come in handy. This project gives children a convenient worksurface and plenty of storage room for supplies. The V-shaped shelving unit allows full access to the pegboard backing, which provides hanging storage. The worksurface has a cutout for easy cleanup of spills. Position a garbage can below the cutout when cleaning up, and leave the cutout covered with the removable cover panel when the worksurface space is needed. We trimmed the top worksurface to curb messy spills.

OVERALL SIZE:
40" HIGH
22" DEEP
34" WIDE

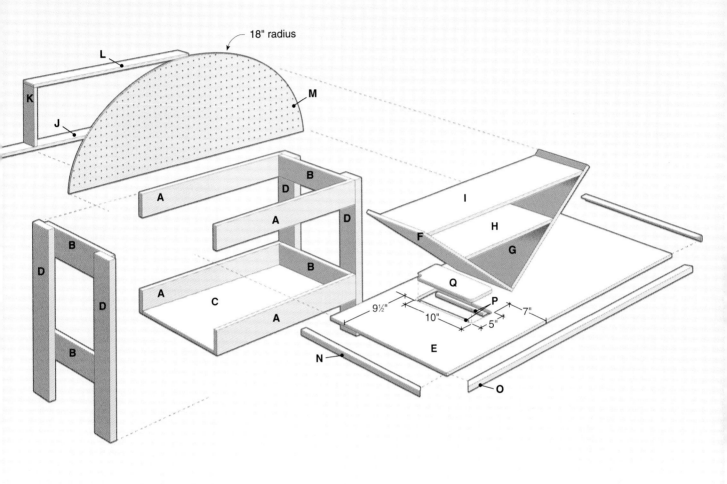

18" radius

L

K

J

M

A

B

D

A

D

B

A

C

A

D

D

D

B

B

N

I

H

F

G

Q

P

E

O

9½"

10"

5"

7"

Cutting List

Key	Part	Dimension	Pcs.	Material
A	Side	¾ × 3½ × 21"	4	Pine
B	End	¾ × 3½ × 16½"	4	Pine
C	Bottom	¾ × 16½ × 19½"	1	Plywood
D	Post	1½ × 3½ × 23¼"	4	Pine
E	Worksurface	¾ × 24 × 36"	1	Plywood
F	Shelf base	¾ × 9¼ × 18"	1	Pine
G	Shelf base	¾ × 9¼ × 17¼"	1	Pine
H	Bottom shelf	¾ × 9¼ × 11¼"	1	Pine
I	Top shelf	¾ × 9¼ × 22½"	1	Pine

Cutting List

Key	Part	Dimension	Pcs.	Material
J	Lower back support	¾ × 1½ × 36"	1	Pine
K	Support side	¾ × 1½ × 11¼"	2	Pine
L	Upper back support	¾ × 1½ × 24"	1	Pine
M	Pegboard	¼ × 18 × 36"	1	Pegboard
N	Side trim	1⅛ × 1⅛ × 22½"	2	Molding
O	Front trim	1⅛ × 1⅛ × 36½"	1	Molding
P	Cleat	¾ × 1½ × 8"	2	Pine
Q	Cover	¾ × 5 × 10"	1	Plywood

Materials: Glue, wood screws (#6 × 1¼", #6 × 2", #6 × 2¼"), 4d finish nails, finishing materials.

Note: Measurements reflect the actual size of dimensional lumber.

Attach the frame with the bottom panels 5" up from the bottoms of the posts.

Attach the worksurface to the top frame in the base unit by driving screws down through the worksurface.

Drive pilot holes through the shelf base pieces and straight into the edges of the shelves.

Directions:
Kiddie Craft Center

BUILD THE BASE. The base unit for the craft center is made from two identical rectangular frames attached inside four posts. Start by cutting the sides (A), ends (B) and bottom (C) to size. Sand all parts with medium-grit sandpaper to smooth out any rough areas after cutting. Position pairs of ends between two sides, and attach them with glue and countersunk #6 × 1¼" wood screws to form two rectangular

frames. Drive the screws through the sides into the ends. Attach the bottom (C) inside one of the frames, using glue and countersunk wood screws. Cut the posts (D) to size. Draw a reference line edge to edge across each post, 5" from one end. Lay a pair of posts on a worksurface, 18" apart. Fasten the frame with the bottom to the posts with glue and counterbored #6 × 2" wood screws **(photo A)**. Make sure the bottom edge of the frame is flush with the line drawn across the posts. The side edges of the posts and frame should be flush. Attach the open frame to the posts so the top frame edges are flush with the top of the post edges. Turn the assembly over and attach the remaining posts on the other side of the frame. Cut the worksurface (E) to size. To make the cleanup cutout, mark a 5 × 10" rectangle on the worksurface, centered 9½" from one end and 7" from one edge (see *Diagram*, page 57). Drill a 1"-dia. starter hole inside one corner of the cutout—the starter hole gives you a finger-grip to re-

move it for cleanup. Insert the jig saw blade and cut along the lines to make the cleanup cutout.

ATTACH THE WORKSURFACE. Cut the cleats (P) to size. Fasten the cleats to the underside of the worksurface with glue and wood screws to create a 1" ledge inside the cutout (the ledge will support the cutout cover). Center the worksurface over the base so it overhangs the back of the base by 1½". Apply glue and drive countersunk #6 × 2½" screws through the worksurface into the posts **(photo B)**.

MAKE THE SHELF ASSEMBLY. The V-shaped shelf assembly is made by attaching the shelves in between the shelf bases (F, G). Begin by cutting the shelf bases to size. Drill pilot holes for countersunk wood screws through the longer base, centered ⅜" from one edge. Apply glue to one end of the shorter base, and attach both base pieces by driving counterbored wood screws through the longer base and into the shorter base. Make sure the shelf base joint forms a right

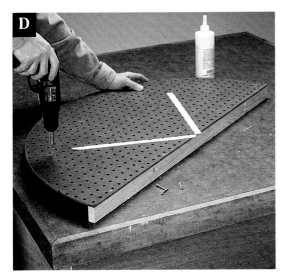

Drive wood screws through the pegboard into the supports and rails.

Position the pegboard and frame onto the shelf assembly.

angle. Set the sole plate of your circular saw so the saw cuts at a 45° angle, and bevel-cut both ends of the bottom shelf (H) and top shelf (I) so the bevels face inward. Position the shelves in the V formed by the shelf bases. Check to be sure the dividers are still connected at a right angle, and outline the shelf positions on the bases. Drive pilot holes through the shelf base pieces and straight into the edges of the shelves **(photo C).** Apply glue to the shelf ends and install them in the shelf base.

ATTACH THE PEGBOARD BACK. Cut the pegboard (M) to size. To cut the pegboard to shape, a semicircle with an 18" radius is cut on the pegboard. To draw the semicircle, drive a ¾" brad at the center of one long edge of the pegboard. Tie an 18"-long piece of string to the brad, and tie the other end around a pencil. With the string taut, draw the 18"-radius semicircle on the pegboard. Cut along the semicircle line with a jig saw. To mark the shelf position on the pegboard, position the shelf assembly with the

tip of the V on the center of the bottom edge. Because it is often hard to see pencil marks on pegboard, apply masking tape to mark the outside edges of the shelf assembly. Cut the lower back support (J), support side (K) and upper back support (L) to size. Fasten the rails to the upper back support at each end with glue and countersunk wood screws driven through the upper back support and into the tops of the support rails. Using a straightedge, draw lines across the top face of the lower support, 6" from each end—drill pilot holes for wood screws centered ⅜" inside those lines. Attach the lower support to the rails by driving countersunk wood screws through the lower support into the rail ends. Apply glue to the edges of the rails and supports, and position the pegboard on them, making sure the bottom edge is flush. Attach the pegboard to the rails and supports with wood screws **(photo D).** Apply glue to the back edges of the shelf assembly. Position the pegboard and support frame onto the shelf

assembly, making sure the outside edges align with the tape **(photo E).** Fasten the parts with wood screws. Attach the shelf and pegboard to the base. Make sure the rear edges of the supports are flush.

APPLY FINISHING TOUCHES. To make the worksurface frame, cut the side trim (N) and front trim (O) to length with a power miter saw **(photo F),** or a miter box and backsaw. Attach the side trim and front trim to the edges of the top with glue and 4d finish nails. Fill all screw holes, nail holes and plywood edges with wood putty. Finish-sand, prime and paint all the surfaces.

Miter-cut corner molding to frame the worksurface.

Dinosaur Rocker

*Who needs a hobbyhorse when this darling dinosaur
is hitched up and ready to ride?*

apture the mystery and excitement that draws children (and a lot of adults) to dinosaurs with this imaginative building project. Cut from a single $4 \times 8'$ sheet of plywood, this dinosaur rocker is a real attention-grabber. The design is fanciful and clever, but it is also engineered for safety. Where most typical hobbyhorses have springs or other moving parts that can easily pinch a wayward finger, this prehistoric fellow rocks on a solid-state platform for a smooth, safe ride.

Although cutting all the contours and shapes does require some patience, there is plenty of room for error built into the shapes, making it an excellent undertaking for the first-time builder as well as for experienced woodworkers. And if you've ever wanted to hone your abilities with a jig saw, this is just the project for you. Most of the patterns for the many interesting shapes and cutouts are provided for you—making them is simply a matter of transferring the shapes accurately to your workpiece, then following the lines.

CONSTRUCTION MATERIALS

Quantity	Lumber
1	¾" × 4 × 8' plywood

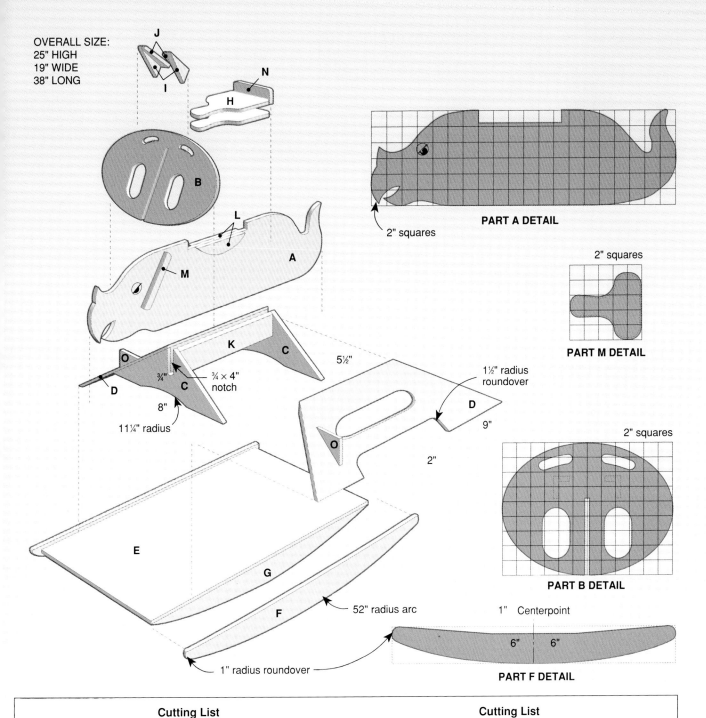

OVERALL SIZE:
25" HIGH
19" WIDE
38" LONG

PART A DETAIL

2" squares

2" squares

PART M DETAIL

PART B DETAIL

2" squares

5½"

1½" radius roundover

9"

2"

¾"
¾ × 4" notch

8"

11¼" radius

52" radius arc

1" radius roundover

1"   Centerpoint

6"   6"

PART F DETAIL

Cutting List				
Key	**Part**	**Dimension**	**Pcs.**	**Material**
A	Body	¾ × 12 × 40"	1	Plywood
B	Frill	¾ × 18 × 24"	1	Plywood
C	Pedestal	¾ × 8 × 20"	2	Plywood
D	Leg	¾ × 14⅜ × 29⅞"	2	Plywood
E	Rocker platform	¾ × 22½ × 37"	1	Plywood
F	Rocker	¾ × 5¼ × 40"	2	Plywood
G	Rocker cleat	¾ × 3½ × 37"	2	Plywood
H	Seat	¾ × 9 × 10"	2	Plywood

Cutting List				
Key	**Part**	**Dimension**	**Pcs.**	**Material**
I	Horn	¾ × 3 × 5"	2	Plywood
J	Horn support	¾ × 1½ × 5"	2	Plywood
K	Pedestal cleat	¾ × 3 × 12"	2	Plywood
L	Seat support	¾ × 2 × 10"	2	Plywood
M	Frill support	¾ × 1½ × 11"	2	Plywood
N	Seat back	¾ × 3 × 4"	1	Plywood
O	Footrest	¾ × 4 × 4"	2	Plywood

Materials: Glue, #6 × 1¼" wood screws, finishing materials.

Note: Measurements reflect the actual size of dimensional lumber.

Make a homemade bar compass from a piece of scrap wood, and use it to draw the curved cutting lines for the rockers and rocker cleats.

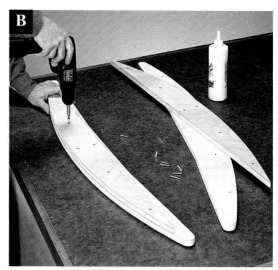

Fasten the rocker cleats to the inside faces of the rockers with glue and wood screws.

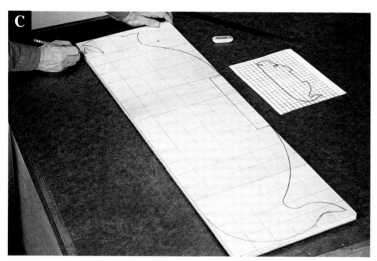

Draw a grid with 2" squares on the plywood, and use the patterns (page 61) as guides for drawing the shapes of the parts.

Directions: Dinosaur Rocker

MAKE THE ROCKERS. If you measure and mark carefully, all the parts for this dinosaur rocker can be cut from one 4 × 8' sheet of plywood. Cut the rockers and rocker cleats first, starting at one end of the plywood sheet, to get off to an efficient start. The arcs on the rockers are just about the only cuts you'll need to make on this project where accuracy is very important. To help you mark smooth, accurate arcs, make a homemade bar compass from a 53"-long piece of straight scrap wood. Simply drill a hole for a pencil with a centerpoint ½" in from one end of the scrap, then drive a 4d finish nail through a point ½" in from the other end. Now, lay out the rough shapes for the rockers (F) and rocker cleats (G) at the top edge of the plywood sheet, and use your bar compass to draw the arc on each piece **(photo A).** When laid out, the rockers should measure 5¼" from the long straight edge to the top of the arc, and the cleats should measure 3½" from the edge to the top of the arc. Measure out from the top of each arc to mark the length of each part, then use an ordinary compass to draw a 1"-radius roundover curve at each end of the rockers (but not the cleats). Cut out the parts with a jig saw, following the cutting lines carefully. Each rocker is flat along the middle of its top edge, but tapers up to the ends. To make these cuts, mark a centerline on the top edge of each rocker, measure straight down 1" from the centerpoint, then draw a 12"-long line that extends 6" on each side of the centerpoint, parallel to the top edge. Use a straightedge to draw a diagonal line from each end of the rocker to the nearer endpoint of the 12" line. Cut along the lines with a jig saw to form the finished top edge of the rocker. Sand out any bumps or irregularities in the cuts. Attach a rocker cleat to the inside face of each rocker, so the curved bottoms are flush and the overhang is equal at each end of the rocker. Use glue and 1¼" wood screws, driven through countersunk pilot holes, to fasten the cleats to the rockers **(photo B).**

Set the sole plate of your circular saw at a 45° angle to make the beveled cuts on the tops and bottoms of the legs.

ATTACH THE ROCKERS TO THE ROCKER PLATFORM. Cut the rocker platform (E). Make sure the platform is square, then set the edges of the platform on top of the rocker cleats to test the fit. The rocker overhang should be equal at the front and the back. Fasten the platform to the tops of the cleats with glue and #6 × 1¼" screws (driven through countersunk pilot holes).

MAKE THE BODY, FRILL & SEAT. The dinosaur body (A), frill (B)—the round cowl that fits behind the head—and the seat pieces (H) are made by transferring the grid patterns from the *Diagram* on page 61 to your sheet of plywood. First, cut workpieces for the parts to the full sizes shown in the *Cutting List*. Draw a 2"-square grid pattern on each workpiece. Then, use the patterns as a reference for drawing the shape onto the workpieces **(photo C)**—it will help if you enlarge the patterns on a photocopier or draw them to a larger scale on a piece of graph paper first. Cut out the shapes carefully with a jig saw, then sand out the rough spots.

Because the seat is made of two identical pieces sandwiched together, lay out and cut one piece, then use it as a template for marking the second piece. Cut out the second piece, then clamp both pieces together and sand them as if they were one workpiece.

MAKE THE PEDESTALS. The pedestals are the main support pieces for the rocker body. They are roughly triangular in shape, with notches at the tops of the triangles where the body fits in. Cut the pedestals (C) to size. Mark a centerline down the middle of each pedestal from top to bottom. Mark guidelines for a ¾"-wide × 4"-long notch, centered on the centerline and starting at the top edge. Use a straightedge to draw a cutting line from each bottom corner to the centerline at the top edge. Mark points 4" out from each side of a centerline on the bottom edge of each pedestal. Draw a line parallel to the bottom edge, ¾" up. Connect the endpoints of the line on the bottom with a smooth arc that reaches the ¾" mark at the top. Extreme accuracy is not crucial for this cut—just try to make sure the arcs are the same on both pedestals. Use a jig saw to cut the diagonal lines from the top of each centerline to the corners, then cut the notches. Finally, cut the arcs in the bottom edges.

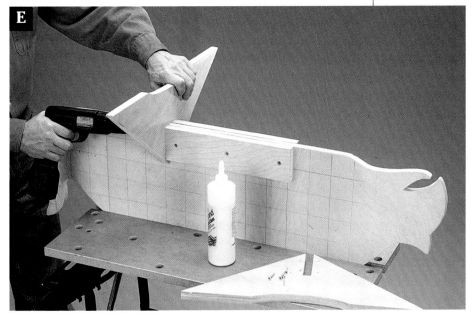

Attach the pedestals to the pedestal cleats that are mounted to the body at the center of the bottom edge.

For any building project that involves extensive cutting and shaping of parts, using good quality sheet goods is extremely important. Laminated plywood with two grade "B" or better faces holds together well and provides a nice surface for painting, but it will require you to do a fair amount of filling of plywood edges with putty before you paint. Solid-core sheet goods offer a good alternative if you don't want to spend a lot of time preparing for painting. The down side, however, is that solid-core sheet goods are more prone to splitting, so you must be more careful when drilling pilot holes and driving fasteners.

MAKE THE LEGS. The legs mount to the top edges of the pedestals to provide stability and footholds for mounting the rocker. The top and bottom edges of the legs are beveled, allowing the parts to butt cleanly against the pedestal edges and the sides of the body. Each side has two 12"-long cutouts for footholds. Start by cutting the legs (D) to size from the plywood. Set the sole plate on your circular saw at a 45° angle to make the bevel cuts. Using a straightedge cutting guide, make the bevel cuts at the bottom and top edges of the legs **(photo D)**. Make sure the resulting bevels are slanted in the same direction on each piece to allow the legs to butt directly against the rocker platform and the body. To cut the legs to shape, mark points 8" from each corner on the top edge. Using a straightedge, draw diagonal cutting lines from these points to the bottom corners. Cut along the lines with a jig saw. Each leg has two cutouts for footholds—one is located at the bottom edge, the top of the other is 5½" down from the top edge. To make the cutout on the bottom edge of each leg, mark points on the bottom edge, 9" from each end. These points mark the ends of the 2 × 12" cutout (see *Diagram*, page 61). Measure up 2" from the bottom and draw a parallel line to mark the top of the bottom cutout. Use a compass set with a 1½" radius to draw roundovers at the ends of the cutouts. Cut along the guidelines to make the bottom cutouts. The top cutouts are 2¾ × 12" in size, 5½" down from the tops of the legs, located directly above the lower cutouts. Draw the cutting guidelines, and use a compass to draw a semicircle with a 1½" radius at each end of the cutouts. Cut along the guidelines with a jig saw to complete the legs.

CUT THE REMAINING PIECES. The dinosaur rocker contains several small parts that are made and attached separately from the main components. Use the cutout pieces from the legs to make the two semicircular seat supports (L)—cut them to size with a jig saw, and sand them to smooth out the edges. Use other pieces of scrap plywood to cut the footrests (O), horns (I), horn supports (J), seat back (N) and pedestal cleats (K) to size. The footrests are easily made by cutting a 4 × 4" square piece of plywood diagonally from corner to corner. Cut the horns to rough size and shape (see *Diagram*), making sure to round the pointed ends with a sander (here is another good place to use gang-cutting and sanding techniques). Simply cut the seat back and frill supports to size, and round over the top corners. The horn supports are basic rectangular cleats, and the pedestal cleats are rectangles with square corners.

ASSEMBLE THE DINOSAUR. With all the parts cut to shape, the dinosaur rocker only needs assembly. Start by fastening the pedestal cleats (K) to the body

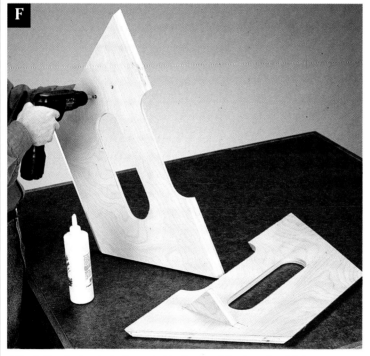

Attach the footrests to the outside faces of the legs by driving screws through the inside faces of the legs and into the ends of the footrests.

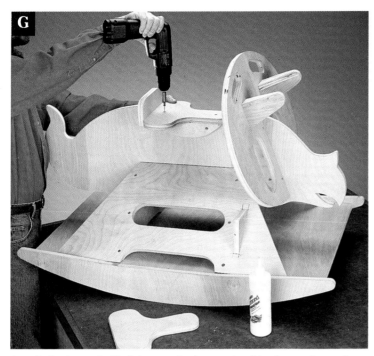

Attach the lower half of the seat in the notch with glue and wood screws, driven through the seat into the seat supports.

(A), flush with the bottom edge and centered end to end. Use glue and countersunk #6 × 1¼" wood screws to attach the parts. Secure the body (A) to your worksurface in an upside-down position, and apply glue to the notches in the pedestals (C). Slide the pedestals over the body at each end of the pedestal cleats (K), and drive counterbored wood screws through the pedestals and into the pedestal cleats **(photo E).** Attach the footrests (O) to the legs with glue and wood screws, so their bottom edges are 2¾" up from the leg bottoms, and their rear faces are 2" from the cutouts on the legs **(photo F).** Stand the assembly upright on its pedestals. Apply glue to the edges of the pedestals (C). Fasten the legs to the top edges of the pedestals with countersunk #6 × 1¼" wood screws, driven through the legs and into the pedestals and body. Center the pedestals on the rocker plat-

form (E), and mark the pedestal positions onto the rocker platform. Remove the dinosaur assembly, and apply glue to the platform, at the pedestal positions. Replace the assembly, and fasten the pedestals to the rocker platform with countersunk wood screws. Fasten the seat supports (L) to the body so their top edges are flush with the square part of the cutout on the top edge of the body. Use glue and 1¼" screws. Attach the horns (I) to the horn supports (J), then attach the horn supports to the front of the frill (B), approximately 5" apart. Position the frill onto the body so that its bottom edge is 6" back from the front of the lower jaw. The back, top edge of the frill should be 15" back from the tip of the nose. This positioning will provide a comfortable riding arrangement. Hold the frill in position, and mark its back edges on the

body. Remove the frill, and attach the frill supports (M) with their front edges along the lines. Apply glue to the front edges of the frill supports, and to the frill slot, then slide the frill onto the body, flat against the frill support edges. Drive countersunk wood screws through the frill and into the frill supports. Center the seat back (N) at the back of the notch in the top of the body, and attach it with glue and wood screws. Position one seat (H) in the notch, and attach it with glue and counterbored wood screws, driven down through the seat and into the seat supports **(photo G).** Apply glue to the top of the fastened seat, and position the unfastened seat on top. Clamp the seat parts, and attach them with countersunk wood screws, driven up through the bottom part of the seat and into the top half.

APPLY THE FINISHING TOUCHES. Sand all the surfaces with medium (100- or 120-grit) sandpaper to smooth out any rough spots, then finish-sand with fine (150- or 180-grit) sandpaper. Prime and paint the rocker—cover all the surfaces with a washable, semigloss enamel paint. We painted the dinosaur a different color than the platform and rocker to add an interesting visual effect. We also painted an eye in the appropriate position on each side of the dinosaur's head. You can enlarge the pattern for the body on page 61, and use it as a reference for painting on eyes of your own—use black model paint to paint the eyes.

Tote Box

*Your child will feel like a grown-up artist or contractor
with his or her own adult-style tote box.*

CONSTRUCTION MATERIALS

Quantity	Lumber
1	1 × 8" × 2' pine
1	1 × 4" × 2' pine
1	1 × 2" × 2' pine
1	½ × ½" × 3' stop molding
1	1"-dia. × 13½" dowel
1	⅜"-dia. × 6" dowel
1	¼" × 1 × 1' hardboard

E very child has a few
cherished possessions
that he or she always
wants close at hand. A tote box
designated for those special
items is a good way to help
your child keep those crayons,
toy tools or building blocks
where they belong. And if your
child has an interest in building
things from wood, this project

is so simple that he or she can
very easily give you a hand
making it.

If you think back to your
high school shop class, you'll
probably remember where
you've seen this design before.
A standard project for a first-
time woodworker, the small
tote or tool box is as simple to
make as it is handy to use.

OVERALL SIZE:
7¼" HIGH
8½" WIDE
12" LONG

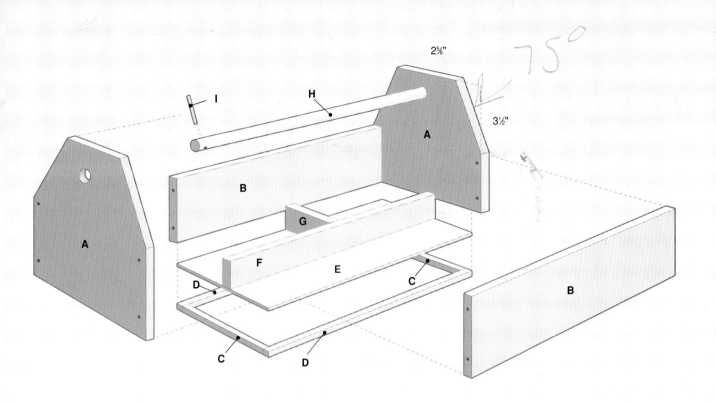

2⅝"

75°

3½"

Cutting List					Cutting List				
Key	**Part**	**Dimension**	**Pcs.**	**Material**	**Key**	**Part**	**Dimension**	**Pcs.**	**Material**
A	End panel	¾ × 7¼ × 8½"	2	Pine	**F**	Long divider	¾ × 1½ × 10½"	1	Pine
B	Side rail	¾ × 3½ × 10½"	2	Pine	**G**	Short divider	¾ × 1½ × 3"	1	Pine
C	End cleat	½ × ½ × 6"	2	Molding	**H**	Handle	1-dia. × 13½"	1	Dowel
D	Side cleat	½ × ½ × 10½"	2	Molding	**I**	Pin	⅛"-dia. × 1½"	2	Dowel
E	Bottom panel	¼ × 5⅞ × 10⅜"	1	Hardboard					

Materials: Wood glue, #8 × 1½" wood screws, 6d finish nails.

Note: Measurements reflect the actual size of dimensional lumber.

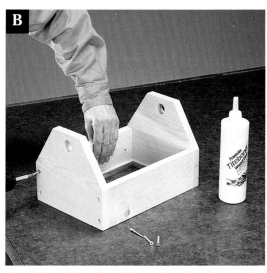

Trim off the top corners of both end panels with a jig saw to create a classic tote appearance.

Attach the side rails to the end panels after attaching cleats to the inside edges of the panels.

Directions: Tote Box

MAKE THE END PANELS. The end panels of the tote box are cut from 1 × 8 pine. Trim off the top corners for classic tote box styling. First, cut the end panels (A). Mark points 2⅝" in from each top corner, on the top edge of each end panel. Also mark points 3½" down from the top corners on the sides of the end panels. Connect the points with a straightedge to mark cutting lines. Trim off both corners at the cutting lines on both end panels, using a jig saw **(photo A).** Find and mark the centerpoint 4¼" in along the top edge of one end panel. Measure down 1⅛" from the centerpoint and mark a point for drilling a 1"-dia. hole for the dowel handle. Stack the end panels together so all the edges line up, and tape the edges to gang them together. Mount a 1"-dia. spade bit in your portable drill, starting with the point of the spade bit on the drilling point. Be careful to keep the bit perpendicular to the workpiece—you may want

to use a portable drill stand. Wrap a piece of medium-grit sandpaper around a 1"-dia. dowel and sand the insides of both holes smooth. This will also widen the hole slightly so the handle can spin freely once it is installed. If the dowel and sandpaper will not fit in the hole, ream out the hole slightly with a file, then try again. Eliminate sharp edges on the parts by sanding with medium sandpaper and a power sander, or use a ⅛" or ¼" roundover bit and a router.

ATTACH THE SIDE RAILS. The side rails are cut from 1 × 4 pine, then equipped with ½"

> **TIP**
>
> *Use a router to make sharp edges smooth. If you do not have much experience operating a router, making roundovers is an easy way to increase your comfort level with the tool. Look for a ⅛" or ¼" piloted roundover bit— piloted bits are designed to follow the edge of a board, with no need for a straightedge to keep them on course.*

square cleats on the inside bottom edges to hold the bottom panel. Cut two side rails (B) from 1 × 4 pine, and cut two end cleats (C) and two side cleats (D) from ½" square stop molding. Round over the edges of the side rails using a sander or a router. Glue a side cleat to each side rail so the bottoms and ends are flush. Glue an end cleat to the inside bottom edge of each end panel. The ends of the end cleats should be 1¼" in from each end of the panel to allow for the thickness of the side rail and the side cleat. After the glue on the cleats dries, apply glue to the end of each side rail, and position the side rails between the end panels so the bottoms and sides are flush. Drill pilot holes for two #8 × 1½" wood screws at each joint. Counterbore the pilot holes deep enough so you can cover the screw heads with wood putty. Drive screws to secure the joints **(photo B).** Check with a framing square to make sure that the sides are square with the ends.

Glue the 1 × 2 pine dividers to the bottom panel, and secure them with 6d finish nails.

TIP

Applying paint or other finishing products to a dowel does not have to be a multi-step process. Simply drive a finish nail through a piece of scrap wood, so the point of the nail sticks out at least ¼". Press one end of the dowel onto the point of the nail so the dowel stands up straight for finishing.

INSTALL THE TRAY BOTTOM. Cut the bottom panel (E) from ¼"-thick hardboard or plywood, using a circular saw and a straightedge cutting guide. Apply glue to the tops of all four cleats. Set the bottom panel onto the cleats and press down firmly. Let the glue dry.

INSTALL THE TRAY DIVIDERS. The tray dividers separate the tote tray into handy compartments. Cut the long divider (F) and the short divider (G) to size from 1 × 2 pine. Round over the sharp edges. Glue the short divider to the bottom panel so it is an equal distance from each end, and one end of the divider is flush with the inside face of one side rail. **(photo C).** Glue the long divider to the tray so the ends are butted against the inside faces of the end panels, and a butt joint is formed with the free end of the short divider. Drive a pair of 6d finish nails through each end panel and into the ends of the long divider. Set the nail heads with a nail set. Also drive a pair of 6d finish nails through the side rail and into

the end of the short divider. Set the nail heads.

FINISH THE TOTE. The tote is finished and painted before the handle is installed. Set all nail heads and fill all counterbores and nail-set holes with wood putty. Sand all surfaces smooth using medium (100- to 120-grit) sandpaper. Then, finish-sand with fine sandpaper (150- to 180-grit). Apply a coat of primer to the tote, then apply at least two coats of enamel latex paint. For extra protection of the finish, apply a coat or two of water-based polyurethane.

INSTALL THE HANDLE. Cut a 1"-dia. pine dowel to 13½" in length to make the handle (H). Sand the ends of the dowel smooth. Secure the handle in a vise or clamp it to your work-surface—be sure to pad the jaws of the vise or the clamps to prevent damage to the dowel. Mark drilling points ½" in from each end of the dowel. Mount a ⅛"-dia. bit in your drill. Hold the drill in an upright position, and drill holes through both drilling points. Cut two 1½"-long pieces from a ⅛"-dia. dowel to make the locking pins

(I) for the handle. Smooth out the ends of the pins. Apply a finish to the handle and the pins. We simply coated the pieces with polyurethane for a protective finish that preserves the natural wood color (which contrasts nicely with the paint we chose for the tote). After the finish on the handle dries, insert the handle through the guide holes in the tops of the end panels. Do not glue the handles. Apply glue to one end of each locking pin. With the handle centered in the guide holes so the overhang is the same on each end, insert a pin into each guide hole to keep the handle from sliding out of the guide holes **(photo D).**

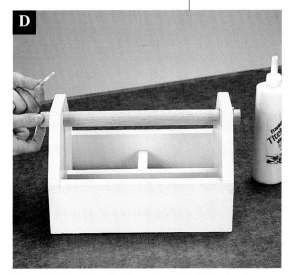

Insert ⅛"-dia. dowels through guide holes in the ends of the handle to keep the handle in place.

Puppet Stage

Encourage your kids to use their imaginations by building this colorful, exciting puppet stage.

CONSTRUCTION MATERIALS

Quantity	Lumber
8	1 × 4" × 8' pine
1	¼" × 4 × 8' lauan plywood
1	¼ × ¾" × 4' screen retainer

Give your child an outlet to express his or her creativity with this simple puppet stage. The curved openings in the top of each stage panel give the puppet stage the feel of an old-time nickelodeon. The square openings below can feature marionettes, or allow the audience to watch their favorite young puppeteer at work.

This puppet stage is made from three hinged panels of lightweight ¼"-thick lauan plywood, so it is easy to fold up for storage, as well as inexpensive to build. To increase the drama of this design, we added velvet curtains behind the stage openings. Just install plain curtain rods on the backs of the panels, then hang any curtains you choose for a backdrop.

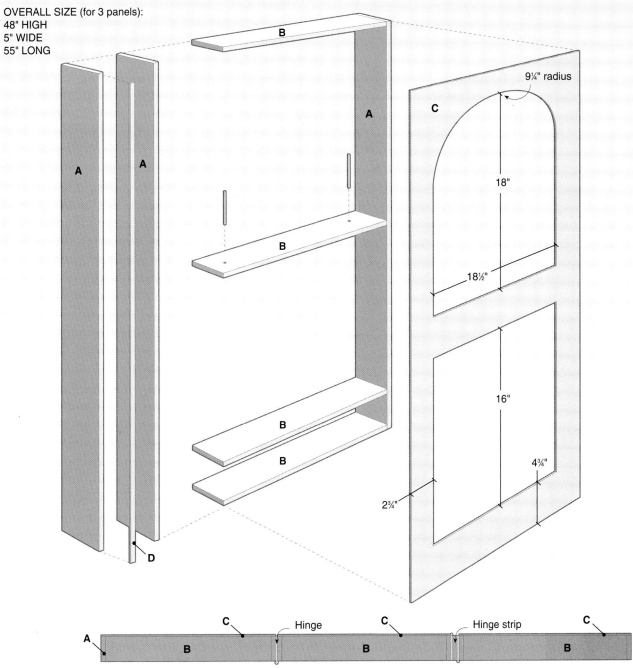

OVERALL SIZE (for 3 panels):
48" HIGH
5" WIDE
55" LONG

B

A

A

A

B

B

B

B

D

9¼" radius

C

18"

18½"

16"

4¾"

2¾"

C   Hinge   C   Hinge strip   C

A

B   B   B

DETAIL FOR HINGE STRIP LOCATION

Cutting List (makes 3 panels)

Key	Part	Dimension	Pcs.	Material
A	Stile	¾ × 3½ × 48"	7	Pine
B	Rail	¾ × 3½ × 22½"	12	Pine
C	Stage panel	¼ × 24 × 48"	3	Plywood
D	Spacer strip	¼ × ¾ × 48"	1	Screen retainer

Materials: Glue, 3 × 3" butt hinges (9), wood screws (#6 × 1¼", #6 × 2"), 1" brads, ⅜"-dia. × 3'-long dowel, finishing materials.

Specialty Items: Compass.

Note: Measurements reflect the actual size of dimensional lumber.

Directions:
Puppet Stage

BUILD THE FRAMES. The puppet stage is made from three box frames made from 1×4 pine boards to support the lauan plywood stage panels. The frames are hinged together so the puppet stage can be folded up for storage. Cut the long stiles (A)—the vertical parts of the frames, and the rails (B)—the horizontal cross pieces which fit between the stiles. Sand all the parts with medium-grit sandpaper to smooth out any rough spots. Set one stile aside; it will be used later as a hinge support plate between panels. Use masking tape to fasten the six remaining stiles together in pairs, with the top and bottom edges flush. Draw reference lines for the crossrail positions across the stile pairs, 4" and 23¾" from the bottom end of each pair **(photo A).** Mark centerpoints for pilot holes ⅜" above each reference line, and ⅜" in from each end of each stile. Remove the tape, and drill pilot holes for #6 × 2" wood screws through the centerpoints. Counterbore the pilot holes so they can be filled with wood putty to conceal the screw heads. Fasten a rail between the tops and between the bottoms of the stiles in each pair, using wood glue and counterbored wood screws.

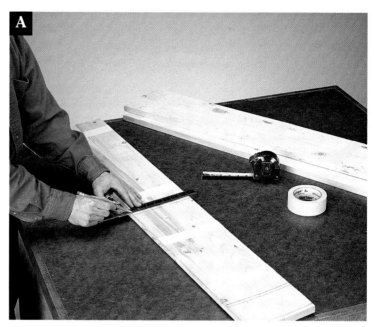

With the stiles taped in pairs, mark the positions of the crossrails.

Also install crossrails between the reference lines.

ATTACH THE FRAMES. Two of the three frames are hinged directly to one another. So the panels fold together, a hinge support strip is needed as a transition between two of the panels. The last crossrail (B) is used to make the hinge support strip. Fasten two frames together with three 3×3" butt hinges attached to the back edges of two frames. Space the hinges so the top and bottom hinges are 1" to 2" from the ends of the frames, and make sure that the barrels of the hinges face toward the back edges of the frames. Use three hinges, spaced the same as the first set, to attach the hinge support strip to the back edge of the third frame. The barrels of the hinges in this set should also face the back edges of the frames. Attach the unhinged side of the hinge support strip to an unattached side of one of the frames you've already hinged together. Use the same spacing, and make sure the barrels in this set of hinges face toward the fronts of the frames **(photo B).** Test to make sure the panels fold together cleanly.

MAKE THE STAGE PANELS. The stage panels are fitted over the fronts of each frame. Openings for the puppets are cut into each stage panel at the top and bottom. Start by cutting the stage panels (C) to size from ¼"-thick lauan plywood. Lay out cutting lines for the panel openings onto one of the stage panels. We cut rounded openings in the tops and square openings in the bottom. To mark the openings, draw a straight line parallel to each side of the panel, 2¾" in from each edge, running all the way from top to bottom. Draw lines 4¾" and 20¾" up from and parallel to the bottom edge. This will create a cutout for the bottom opening. Draw a line across the face of the panel, 21" down from the top. Mark the center of the panel on the

Attach a hinge support strip as a transition piece, using butt hinges, when hinging together two of the three frames.

Gang-cut both openings into all three panels at the same time, using a jig saw.

Attach the stage panels to the fronts of the frames, making sure the space between panel openings falls over the middle rail.

top edge, and draw a line (at least 15" long) straight down from the centerpoint. Mark a point on the centerline, 12¼" down from the top edge. Tack a finish nail at this point, and tie a string to the nail. Tie a pencil to the other end of the string, 9¼" away from the nail. Pull the string taut, and draw a semircircle that intersects the lines at the sides of the panel, with the top of the semicircle

about 3" down from the top edge of the panel. Clamp the other two stage panels below the marked panels, and make the cutouts in all three panels at once, using a jig saw inserted into a starter hole for each opening **(photo C).**

ATTACH THE PANELS. Sand the edges of the panels and cutouts smooth, then tack each panel to the front of a frame,

using 1" brads **(photo D).** Use a nail set to set the nails. If the edges of the panels overhang the frame, cut or sand them until flush.

APPLY FINISHING TOUCHES. Cut a piece of ¼ × ¾" screen retainer molding to make the spacer strip (D). Attach the strip to the front of the hinge support strip as a spacer, using 1" brads, so the strip is flush with the faces of the stage panels when the stage is opened up. Drill ⅜"-dia. holes for dowels in the corners of the middle stiles. These dowels are used as spindles for holding puppets or marionettes. Cut a ⅜"-dia. dowel to 4" lengths. Glue the dowel ends, and insert them into the holes in the rails. Prime the surfaces, then paint the puppet stage (we used semigloss latex enamel paint). You may also hang curtains behind the openings in the stage panels (see page 70).

Bedroom Divider

If you have two kids sharing a room, use this bedroom divider to define boundaries and add storage space.

PROJECT
POWER TOOLS

Perfect for an expanding family, this bedroom divider solves the space problems that arise when two children share a room. The symmetrical design offers identical storage to both children, eliminating space disputes.

The divider is made from two separate units—the closet unit and the shelf unit. The shelf unit features adjustable bookshelves and plenty of flat surfaces for hanging artwork or notes. The closet unit features Shaker pegs in the back for hanging jackets, sweaters or book bags. This bedroom divider is symmetrical, so when the shelf unit is pushed against the middle of the closet unit, two identical sections are formed. With the rails along the bottom of the closet unit creating a shallow bin, it also makes a great place to store shoes.

You can attach the two units with wood screws where they meet to form a single structure, or keep them separate to make sure they can be easily moved. If the space in the childrens' bedroom is truly cramped, just one divider may be enough. With the closet unit positioned against the wall, the divider is open and accessible. If you position the shelf unit against the wall, it becomes more private and enclosed. Whether you position the shelf or the closet unit against the wall, the bedroom divider can help you get the most from a single room.

CONSTRUCTION MATERIALS

Quantity	Lumber
2	¾" × 4 × 8' plywood
1	¼" × 4 × 8' plywood
5	¾ × ¾" × 8' quarter-round

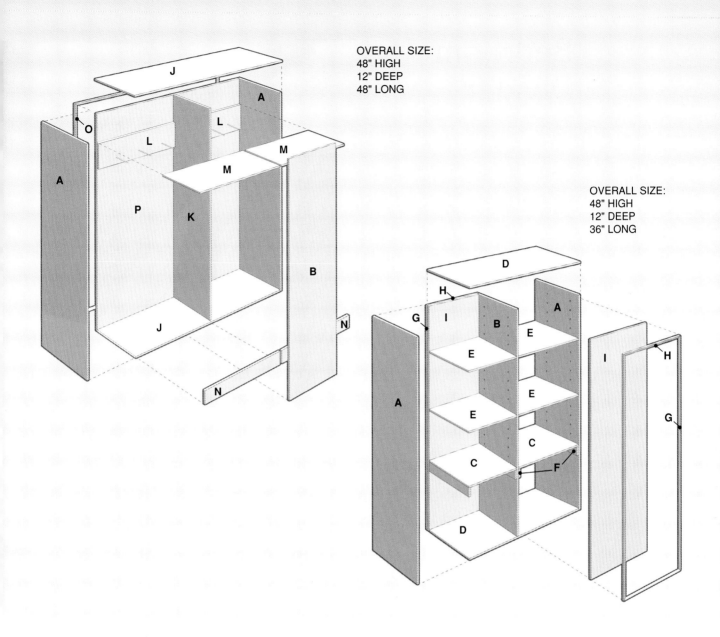

OVERALL SIZE:
48" HIGH
12" DEEP
48" LONG

OVERALL SIZE:
48" HIGH
12" DEEP
36" LONG

	Cutting List			
Key	**Part**	**Dimension**	**Pcs.**	**Material**
A	End	¾ × 12 × 48"	4	Plywood
B	Closet divider	¾ × 12 × 46½"	2	Plywood
C	Fixed shelf	¾ × 11 × 16⅞"	2	Plywood
D	Case cap	¾ × 12 × 34½"	2	Plywood
E	Bookshelf	¾ × 10¾ × 16¾"	4	Plywood
F	Shelf cleat	¾ × 1½ × 9½"	4	Plywood
G	Frame stile	¾ × ¾ × 46½"	4	Molding
H	Frame rail	¾ × ¾ × 16⅞"	4	Molding

	Cutting List			
Key	**Part**	**Dimension**	**Pcs.**	**Material**
I	Case back	¼ × 16⅞ × 46½"	2	Plywood
J	Closet cap	¾ × 12 × 46½"	2	Plywood
K	Partition	¾ × 10¼ × 46½"	1	Plywood
L	Stretcher	¾ × 6 × 46½"	1	Plywood
M	Closet shelf	¾ × 10¼ × 22⅞"	2	Plywood
N	Closet rail	¾ × 4 × 22⅞"	2	Plywood
O	Closet trim	¾ × ¾ × 46½"	4	Molding
P	Closet back	¼ × 46½ × 46½"	1	Plywood

Materials: Glue, ½"-dia. Shaker pegs (6), wood screws (#4 × ¾", #6 × 1½", #6 × 2"), 4d finish nails, finishing materials.

Specialty Items: ¼" pegboard for drilling guide.

Note: Measurements reflect the actual size of dimensional lumber.

Directions:
Bedroom Divider

BUILD THE SHELF UNIT CABINET. The shelf unit is the smaller of the two sections. It features adjustable bookshelves and flat, plywood backs that make great bulletin boards or display areas. The shelf unit frame is built first, and the shelves are added later. The frame consists of the ends (A), case caps (D), divider (B) and case back (I). The case caps are positioned between the ends with their edges flush and attached with wood screws and glue to house the adjustable shelf assembly. Throughout all the construction steps, you should check to make sure the frame is square. Countersink pilot holes for screws deep enough so they can be filled with wood putty to cover the screw heads. Start by cutting the ends (A) and dividers (B). Sand all parts with medium-grit sandpaper to smooth out the rough spots after cutting. Set aside one pair of ends and one divider for later assembly on the closet unit. Use a square to draw guidelines on the faces of the remaining two ends and divider—12" from the bottom of the ends and 11¼" from the bottom of the divider. These lines mark the position of the shelf cleats (F), which support

TIP

When checking a cabinet for square, measure diagonally from corner to corner. If the measurements are equal, the cabinet is square. Apply pressure to one side or the other with your hand or clamps to push a cabinet back into square.

Use a 12 × 36"-long piece of ¼" pegboard to guide your drilling as you make holes for shelf pins in the ends and divider.

the fixed shelf (C). Use a 12 × 36"-long piece of pegboard as a guide to drill ⅜"-deep holes for shelf pins in the divider and ends. Start the holes 4" above the guidelines and 2" from the front and back end edges **(photo A).** Drilling a hole every 1" will give you plenty of options for varying shelf heights. Place tape on the pegboard to mark the rows you use. Marking the pegboard helps to ensure that the holes are drilled in the same row positions. Stagger the holes on opposite sides of the divider faces to prevent accidentally connecting the holes when you drill them. Sand the faces when the drilling is complete. Cut the case caps (D) to size from ¾"-thick plywood. Use glue and counterbored #6 × 2" wood screws to fasten the caps be-

tween the ends so that their edges are flush. Drive the screws through the ends into the cap edges. Measure diagonally from corner to corner to check to make sure the frame is square. Mark 16⅞" from each end on the inside faces of the case caps. Glue the top and bottom of the divider and install it between these lines, making sure the edges are flush. Drive counterbored wood screws through the caps into the divider ends to secure the divider in the frame.

ATTACH THE CASE BACKS. To complete the shelf unit, you need to attach the case backs (I), trim, and bookshelves (E). The divider bisects the frame into two sections—each section has a case back on one

Attach the frame stile and frame rail to the ends, divider and case caps with glue and finish nails.

Attach the fixed shelves in the bookshelf unit to the cleats mounted to the sides and the divider.

side. Make sure the case backs are on opposite sides of the frame, so the shelving will be exposed on opposing sides. For more information on the positioning of these parts, see the *Diagram* on page 75. Use a miter box to miter-cut the frame stile (G) and frame rail (H) from quarter-round molding. The trim is attached to the frame to secure the case backs in place. Use glue and 4d finish nails to attach the trim on opposite sides and ends of the shelf unit frame, flush with the divider, case caps and ends **(photo B).** Cut the case backs to size. Measure diagonally from corner to corner on the frame to check for square. If the measurements are different, apply pressure to the frame on a corner until it is square. Push the case backs into the

frame on both sides of the divider until they butt against the trim. Fasten the case backs to the trim with countersunk #4 × ¾" wood screws, driven through the case backs and into the trim pieces. It is important to use screws to fasten the case backs to the trim. The pounding of a hammer would dislodge the trim from the shelf unit frame. Cut the shelf cleats (F) to size, and fasten them on the guidelines drawn on the ends and divider using glue and counterbored #6 × 1¼" wood screws. When installed, the shelf cleats (and shelves) should butt against the case back.

INSTALL THE BOOKCASE SHELVES. In addition to providing valuable shelf space, the fixed shelves add to the struc-

tural stability of the shelf unit, and help to keep the frame square. Cut the fixed shelves to size. Glue the tops of the shelf cleats, and drop the fixed shelves in place on the cleats. Drive counterbored wood screws through the top of the fixed shelves and into the shelf cleat tops **(photo C).** Cut the bookshelves (E) to size from ¾"-thick plywood. Insert removable

Use wood putty to fill exposed plywood edges and screw holes in the wood. Sand smooth before finishing.

Cut the notch for the stretcher on the back edge of the partition using a jig saw.

¼" shelf pins into the holes on the divider and ends at desired intervals. Position the shelves into the shelf unit. You will want to remove the adjustable shelves before painting the bedroom divider.

APPLY FINISHING TOUCHES. Fill any surface irregularities with wood putty **(photo D),** and finish-sand the shelf unit with fine-grit sandpaper to smooth out any rough surfaces. Apply a flat latex paint to the shelf unit; use a light brush load around the shelf holes to avoid filling them with paint. Once the paint has dried, apply two coats of polyurethane to protect the finish.

MAKE THE CLOSET UNIT FRAME. The closet unit is built in much the same way as the shelf unit. Unlike the shelf unit, the closet unit is open on one side only and has no adjustable shelves. Two ends (A), closet caps (J) and closet back (P) are used to make the closet unit frame. Cut the closet caps (J) to size. Attach the closet caps between the ends with glue and wood screws, making sure their edges are flush. Use a miter box to cut the closet trim (O) to size. Use glue and 4d finish nails to attach the trim flush with the edges on one side of the ends and closet caps. The trim will frame the closet back (P) and hold it in place. Cut the closet back to size, and slide it into the frame so that it butts against the back edges of the closet trim. Attach the closet back by driving #4 × ¾" wood screws through the back into the trim.

COMPLETE THE CLOSET UNIT. With the end unit frame assembled, the partition (K), stretcher (L), closet rail (N) and divider (B) must now be attached. Cut these parts to size, and sand their edges with medium-grit sandpaper. Carefully mark the stretcher position on the partition before attaching the parts. (In addition to adding structural support, the stretcher has Shaker pegs attached to it for hanging jackets or sweaters.) The partition has a notch cut into its back edge to hold the stretcher. Mark a ¾"-wide × 6"-long notch on the partition—starting 32½" from the bottom of the partition. Cut the notch with a jig saw **(photo E).** Center the stretcher in the notch. Use a square to make sure the stretcher and partition are square, and attach them with glue and countersunk #6 × 2" wood screws, driven through the stretcher and into the back of the partition. Glue their edges, and slide the partition and stretcher into the frame so that they butt against the closet back. Fasten them with counterbored #6 × 2" wood screws driven through the closet caps and ends and into their edges. Center the divider in front of the partition, between the closet caps. Drive counterbored #6 × 2" wood screws through the closet caps into the top and bottom of the divider, and through the divider into the partition to attach it. Position the closet rails between the divider and ends on the

front edge of the bottom closet cap. Glue the edges, and drive wood screws through the ends into the closet rails, and through the closet rails into the divider. Stand the end unit up, and attach the closet shelves to the ends and divider with counterbored wood screws and glue **(photo F).**

APPLY FINISHING TOUCHES. Begin by drilling three evenly spaced ¾"-deep × ½"-dia. holes into each stretcher for Shaker pegs **(photo G).** Glue their ends, and insert the Shaker pegs into the holes. Fill all edge irregularities and exposed screw holes and nail holes with wood putty. Finish-sand all the surfaces to smooth out the cor-

ners and rough edges. Paint the closet unit to match the shelf unit.

ARRANGE THE DIVIDER. Position the divider against a bedroom wall, forming a T-shape. With the shelf unit positioned against the wall, you create a more private divider, suitable as dressing rooms. With the closet section against the wall, the units become more open and accessible. For increased stability, fasten the shelf unit to the closet unit with wood screws, driven through the shelf unit end and into the closet divider. For easy disassembly, use L-brackets to attach the shelf and closet units.

Drive wood screws through the divider and into the closet shelf edges to secure them in the frame.

Drill evenly spaced holes in each stretcher for the Shaker pegs.

Bookworm Study Set

*A matching lamp base and bookends
make studying fun for young learners.*

CONSTRUCTION MATERIALS

Quantity	Lumber
1	¼" × 2 × 4' plywood
1	¾" × 2 × 4' plywood
1	1 × 2" × 6' pine

The engaging cartoon bookworm that weaves its way through this lamp base and matching set of bookends will help your kids look forward to study time. Along with sending the positive message that learning is fun, this study set also fulfills a valuable, practical function by bringing additional light and storage to your child's bedroom.

This study set project is a learning experience for the parent or grandparent who builds it, as well. Equal parts craft and carpentry, cutting the intricate bookworm shapes and applying the decorative painted finish will teach you a lot about your own level of enthusiasm for detail-oriented crafts. If you enjoy it, you may have discovered a new hobby that could lead quickly to more intricate scrollsaw or bandsaw projects. But even if you don't particularly enjoy the process, you'll still have a special gift to give to a special child or grandchild.

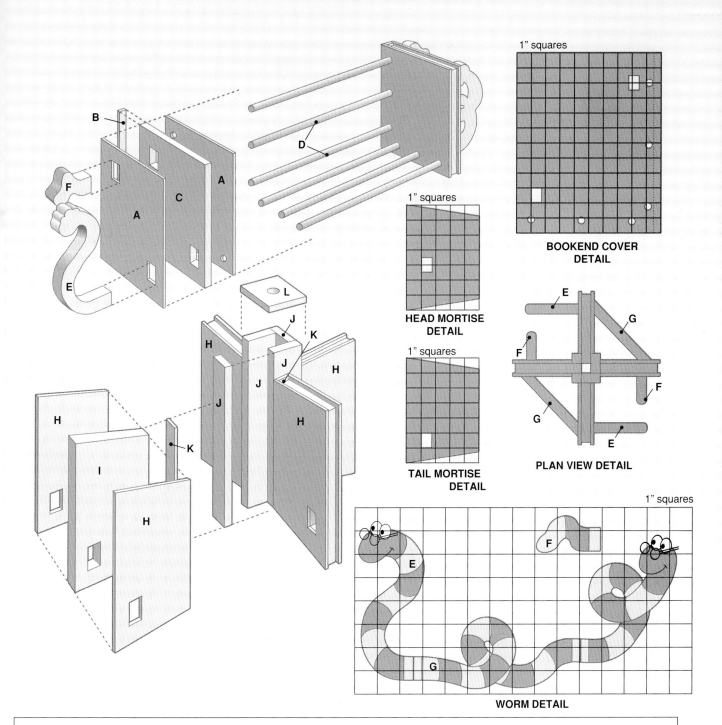

1" squares

BOOKEND COVER DETAIL

1" squares

HEAD MORTISE DETAIL

1" squares

TAIL MORTISE DETAIL

PLAN VIEW DETAIL

1" squares

WORM DETAIL

Bookends Cutting List				
Key	**Part**	**Dimension**	**Pcs.**	**Material**
A	Bookend cover	¼ × 10 × 12"	4	Plywood
B	Spine	¼ × ¾ × 12"	2	Plywood
C	Page	¾ × 9½ × 11¾"	2	Plywood
D	Dowel	½ × 19½"	6	Dowel
E	Worm head	¾ × 5½ × 7"	4	Plywood
F	Worm tail	¾ × 2½ × 3"	4	Plywood

Lamp Base Cutting List				
Key	**Part**	**Dimension**	**Pcs.**	**Material**
G	Worm middle	¾ × 4 × 7½"	2	Plywood
H	Lamp cover	¼ × 5 × 8"	8	Plywood
I	Lamp page	¾ × 4½ × 7¾"	4	Plywood
J	Post	¾ × 1½ × 11¼"	4	Pine
K	Lamp spine	¼ × ¾ × 6½"	4	Plywood
L	Post cap	¾ × 3½ × 3½"	1	Plywood

Materials: Glue, epoxy, #4 × 1¼" wood screws, 1" wire brads, ½"-dia. dowels, lamp hardware kit with a 12"+ tube, finishing materials.

Specialty items: Flat wood file, fine scrolling jig saw blade, coping saw.

Note: Measurements reflect the actual size of dimensional lumber.

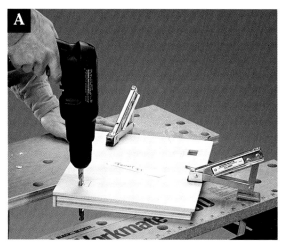

Drill a starter hole, then cut the mortises for the bookworm with a jig saw.

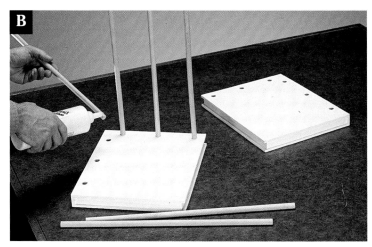

Prime or prime and paint the bookends, then glue dowels into the dowel holes to connect the bookends.

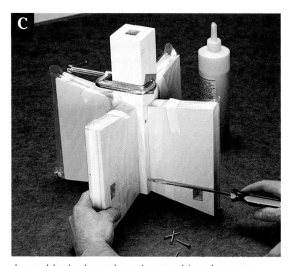

Assemble the lamp base by attaching the posts.

Directions:
Bookworm Study Set

BUILD THE BOOKENDS. The bookends are made by building three-part "book" assemblies from plywood and connecting them with dowels. Cut the bookend covers (A), spines (B) and pages (C) to size. Sand all parts smooth. Use glue and 1" wire brads to attach the spines to one edge of each page. Center the spines on the pages so ⅛" of the spine extends beyond the top and bottom of each page. When the spines are attached, designate the inside and outside covers,

and mark them accordingly. Each outside cover contains two ¾"-wide × 1"-high mortises to hold the ends of the bookworm parts. Mark the mortise locations (see *Diagram*, page 81) on the outside covers, then clamp each outside cover to a page and spine. Make sure the covers extend past the pages ¼" on the front and ⅛" on the top and bottom edge. Place the clamped assembly, page down, on a worksurface, and drill starter holes for the mortises, through both the outside covers and the pages **(photo A).** Cut the mortises into both parts with a jig saw. With the covers and pages still clamped together, use a flat wood file to file the edges of the mortises so they are square. Clamp the inside covers together, and mark centers for the dowels (see *Diagram*, page 81). Note that the dowels on the bottom edge form a line that slopes downward slightly, and the dowels on the back edge of the bookends tilt back slightly from bottom to top (this helps prevent books from falling out of the bookends). Drill a ½"-dia. hole at each centerpoint. Now, unclamp the inside covers,

and position each inside cover over a page, maintaining the same ⅛" and ¼" setbacks established when cutting the mortises. Be careful not to confuse the front and back of the inside covers, then clamp each cover to a page and extend the ½"-dia. holes all the way through the page. Unclamp the parts and wipe the surfaces clean. Now, assemble the two books by sandwiching a page between each pair of covers, making sure the mortises and dowel holes are aligned. Bond the books together with glue and 1" wire brads. Set the heads of the brads and fill the holes and the exposed plywood edges with wood putty. Sand all surfaces, then apply a heavy layer of white primer to each book. Cut the dowels (D) to size, apply glue to the ends, and insert them into the dowel holes **(photo B)**. After the glue dries, mask the exposed page edges with masking tape, then paint the covers with enamel paint (we decided to leave the dowels with a natural color, so we also masked the ends of the dowels where they meet the covers before painting).

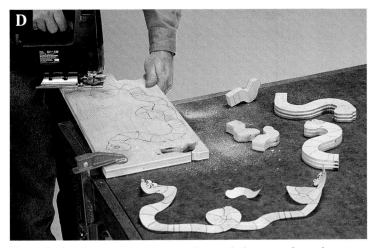

Trace the bookworm pattern onto plywood, then cut along the cutting line with a jig saw fitted with a fine scrolling blade.

MAKE THE LAMP BASE. The lamp is also designed with the book-and-bookworm motif. The base uses four smaller books that are foreshortened from front to back. The spine of each book is attached to a 1 × 2 post, then the post sections are joined together to form a hollow post that allows room for the lamp kit. Cut the lamp covers (H), lamp pages (I) and lamp spines (K) to size. Trim the tops and bottoms of the lamp covers and pages so they slope toward the center of the book ¾" from the front to the back (see *Diagram*). Attach the lamp spines to the lamp pages with glue and 1" wire brads. Assemble pairs of covers with pages the same way as with the bookend covers. Cut a mortise all the way through each "book" according to the positioning shown in the *Diagram:* cut two books with head mortises, and two with tail mortises. Make sure the covers with the head mortises are opposite one another when the base is assembled. Cut the posts (J) and post cap (L) to size. Attach a post to the spine of each book, with glue and #4 × 1¼" screws driven through

the posts and into the spines. The bottoms of the spines and posts should be flush, and the inside edge of each post should be ½" from the left side of the book (when the book is standing upright). Finish-sand each book assembly, and apply primer and paint—we suggest you use the same color schemes used for the bookends. After the paint dries, tape plastic over the nonmating surfaces of the assemblies for scratch protection. Apply glue to the posts, and secure them together into a hollow lamp post **(photo C)** with #4 × 1¼" screws (see *Diagram*). Remove plastic protection. Drill a ⅜"-dia. hole through the center of the post cap for a threaded lamp tube (check the diameter of the tube in your lamp kit, and make sure this hole matches it). Center the cap over the posts and fasten it with glue and wire brads.

MAKE THE BOOKWORMS. The easiest way to draw the serpentine shapes for the bookworms is to transfer the grid pattern from the *Diagram* onto a piece of graph paper to full size, then cut out the shape in the paper

to form a template. Lay the template over a piece of ¾" plywood, and trace around it. Mark cutting lines (the worm will be divided into sections) and draw guidelines for painting, then cut out the shape with a jig saw and fine scrolling blade **(photo D).** Drill holes in the internal cutouts, and cut them to shape with a coping saw. Cut worm sections for both the bookends and the lamp base (For the lamp base, miter-cut the worm middle ends at 45° angles). Sand the edges smooth, fill with putty, finish-sand, then paint the bookworm (we followed the cartoon pattern, using craft paint in the colors shown).

APPLY FINISHING TOUCHES. Glue the bookworm heads and tails into the mortises in the bookends and lamp base, following the *Diagram*. Tack the bookworm middle sections in the lamp base in place with quick-setting epoxy, then secure each end of the middle sections with a 1" wire brad driven through a pilot hole **(photo E).** Install the lamp hardware kit according to the manufacturer's directions.

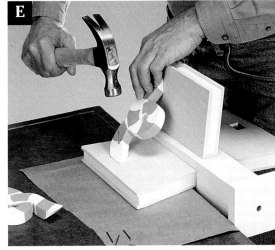

Drive wire brads through the worm middle and into the lamp covers.

Action Figure Modules

Imagination runs wild when you build these fun accessories for your child's collection of dolls and action figures.

Kids today have entered the age of the action figure—those plastic or rubber scaled-down versions of athletes, fashion models, or television and movie stars. Made with joints that flex and equipped with an arsenal of accessories, action figures need a place to roam, and that's where these action figure modules come into play. The modules shown above are basically three plywood boxes of differing sizes and shapes. They are open in front, and they have holes cut in the sides to create passageways between modules for action figures that are always on the go. These modules are simple and inexpensive to make, but they will create endless hours of fun for your children by giving their favorite toys a creative landscape that is all their own.

CONSTRUCTION MATERIALS

Quantity	Lumber
1	¼" × 4 × 8' plywood
1	¾" × 2 × 4' plywood
1	½" × 8' quarter-round molding

OVERALL SIZE (when arranged as shown):
30" HIGH
10" WIDE
25" LONG

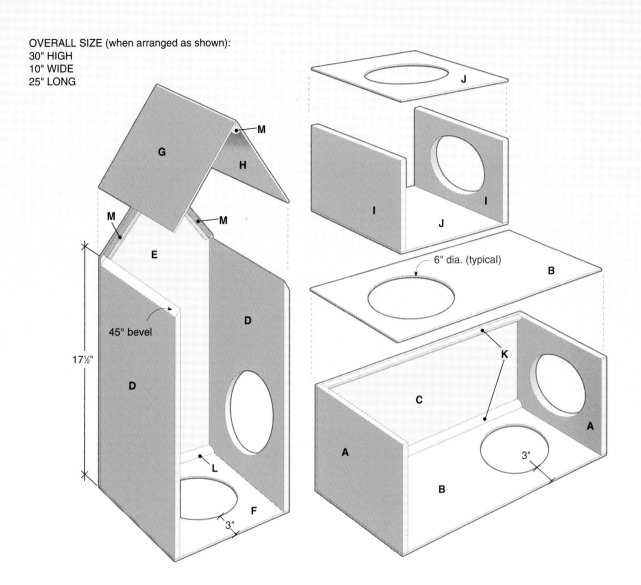

6" dia. (typical)

45° bevel

17½"

3"

Cutting List

Key	Part	Dimension	Pcs.	Material
A	Main side	¾ × 10 × 12"	2	Plywood
B	Main cap	¼ × 12 × 24"	2	Plywood
C	Main back	¼ × 10½ × 24"	1	Plywood
D	Tower side	¾ × 12 × 18"	2	Plywood
E	Tower back	¼ × 13½ × 24¼"	1	Plywood
F	Tower floor	¼ × 12 × 13½"	1	Plywood
G	Short roof	¼ × 9¾ × 12¼"	1	Plywood

Cutting List

Key	Part	Dimension	Pcs.	Material
H	Long roof	¼ × 10 × 12¼"	1	Plywood
I	Insert side	¾ × 9⅜ × 11½"	2	Plywood
J	Insert cap	¼ × 11½ × 11½"	2	Plywood
K	Main cleat	½ × ½ × 22½"	2	Molding
L	Tower cleat	½ × ½ × 12"	1	Molding
M	Ridge pole	½ × ½ × 8"	3	Molding

Materials: Glue, brads (½", ¾"), finishing materials.

Note: Measurements reflect the actual size of dimensional lumber.

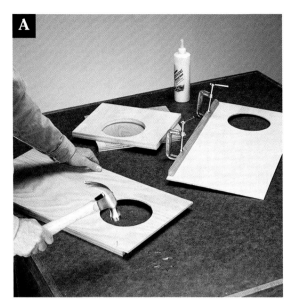

Tack quarter-round molding cleats at selected joints between panels (see Diagram, page 85).

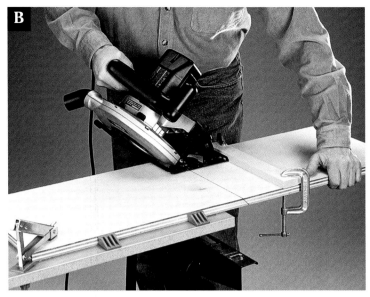

Bevel-cut the tops of the tower sides, using a circular saw with the blade set at a 45° angle.

Directions:
Action Figure Modules

BUILD THE MAIN MODULE. The main module is a larger rectangular unit, spacious enough to hold the smaller insert module and to serve as a base for the tower. Like the other modules, it features passageways that are cut into the parts with a jig saw. You can position the holes any way you want, but try to arrange the passageways so they line up from module to module, allowing the action figures to "move" from one module to another. Start by cutting the main sides (A), main caps (B) and main back (C) to size. Sand all the parts with medium-grit sandpaper to smooth out any rough

TIP

Use a router and template to cut many larger holes of the same size. Simply cut a hole of the correct size into a piece of plywood, using a jig saw, to make the template. Position the template over the wood to be cut, and cut along the edges of the template with a piloted straight bit for template cutting (drill a starter hole first).

edges after cutting. We cut passageways into the main sides, caps and back. First, draw 6"-dia. circles with a compass (or find a coffee can with a diameter of about 6" and trace around the base) on the parts. Locate the circles so the edges are 3" in from the edges of the workpiece (this will put the centerpoint of each circle 6" from the edges). After you have drawn the circles, drill a ⅜"-dia. starter hole inside each circle. Insert your jig saw blade into the hole, and cut along the cutting lines to make the passageways. Sand the edges smooth. Cut the main cleats (K) to size from ½" quarter-round molding. Position each cleat on the inside face of a main cap, so a square edge is flush with the back edge of the cap and the ends of each cleat are ¾" in from the side edges of the caps. Attach the cleats at these locations, using glue and ½"-long wire brads **(photo A).** Clamp the cleats before driving the brads to keep them from slipping out of position. Fasten the

main cap to the main sides with glue and ¾" wire brads driven through the caps and into the top and bottom edges of the sides. Finally, attach the main back to the back edges of the main cleats with glue and ½" brads. Also attach the back to the back edges of the sides, using glue and ¾" brads. Carefully set the heads of the brads with a nail set, and fill all the nail holes with wood putty.

BUILD THE TOWER. The tower module has a pitched roof that is held together by a ridge pole made from ½" quarter-round molding. Start by cutting the tower sides (D) to size. Set the sole plate on your circular saw to 45° to bevel-cut the tops of the tower sides so they form a right angle when butted together **(photo B).** Assemble the unit. Cut the tower back (E), tower floor (F), short roof (G) and long roof (H). Draw 6"-dia. circles on the tower floor and tower side to make passageways (see *Build the main module*). Use the holes in the main module as guides to

C

Position a piece of scrap in the tower to support the ridge pole as you attach the long roof section.

make sure the holes are aligned and of the same size. To cut the tower back to shape, mark the center of the tower back on its top edge. Also mark the side edges of the tower back, 17½" up from the bottom. Draw diagonal cutting lines from the top-edge centerpoint to the side marks, and cut along the lines with a jig saw or circular saw. Sand the cuts smooth. Use glue and ¾" brads to fasten the tower back to the tower sides, making sure the outside edges are flush and the bevels on the tower sides align with the slanted top edges of the tower back. The bottom edge of the tower back should extend ¼" below the bottoms of the tower sides. Cut the tower cleat (L) to length from ½" quarter-round molding. Position the cleat flush with the back edge of the tower bottom, ¾" in from each side edge. Attach the tower cleat to the tower bottom with glue and ½" brads, driven through the tower bottom and into the tower cleat. Cut the ridge pole (M) to

length and attach it at the inside face of the short roof section, so the ends of the ridge pole are flush with the ends of the roof section, and a square edge of the pole is flush with the top edge of the roof section. Use glue and ½" brads. Now, set the short roof section onto the top of the tower, so the ridge pole aligns with the gable peak in the tower back. Apply glue to the back end of the ridge pole, to the back edge of the roof section and to the beveled edge of the tower side, then drive a ¾" brad through the peak of the tower back and into the end of the ridge pole. Cut a piece of scrap wood to 29½", and insert it underneath the ridge pole for temporary support. Drive ¾" brads down through the short roof section and into the top of the tower side. Attach the short roof to the tower top with glue and ½" brads. Apply glue to the free side of the ridge pole and to the top edges of the short roof, tower back and free tower side. Attach the long roof sec-

tion with ½" brads driven into the ridge pole and ¾" brads driven into the side **(photo C).** Set the brads with a nail set.

Build the insert module. The insert module is simply a box with 6"-dia. holes cut into the side and top. Cut the insert sides (I) and insert caps to size. Cut 6"-dia holes in one side and in the top cap (see *Build the main module*). Make sure the holes align with matching holes in the other modules. Fasten the insert caps to the sides with glue and ¾" brads. Set the heads of the brads.

Apply finishing touches. Fill all the nail holes with wood putty. Finish-sand the surfaces with fine-grit sandpaper. Apply enamel paint to the modules. Be creative with the paint application, if you like: paint the modules different colors, or use decorative painting techniques. Another nice touch is to glue appliqués—we used scraps of wallpaper—to the sides of the modules **(photo D).**

D

Glue decorative appliqués, such as wallpaper scraps, to the modules for a personalized touch.

Treasure Chest

*Keep cherished toys and prized belongings in a safe place
with this beaded pine treasure chest.*

CONSTRUCTION MATERIALS

Quantity	Lumber
6	1 × 3" × 8' pine
1	½ × 1¼" × 8' pine stop molding
1	¾" × 2 × 4' AB plywood
7	¼ × 3" × 8' beaded paneling

Treasures are in the eyes of the beholder. One child's gold dubloons are another child's old bottle caps. Whatever perceived valuables your kids may have, they'll enjoy the opportunity to keep them safe and secure in this trusty treasure chest. Built entirely of pine and beaded pine paneling, this chest incorpo-

rates simple construction designs that give the appearance of more complicated counterparts. The lid is held open by a lid support and, when closed, is kept structurally sound by the side and end lips. The treasure chest is suitable for either a natural or painted finish and can be accented with decorative trunk hardware.

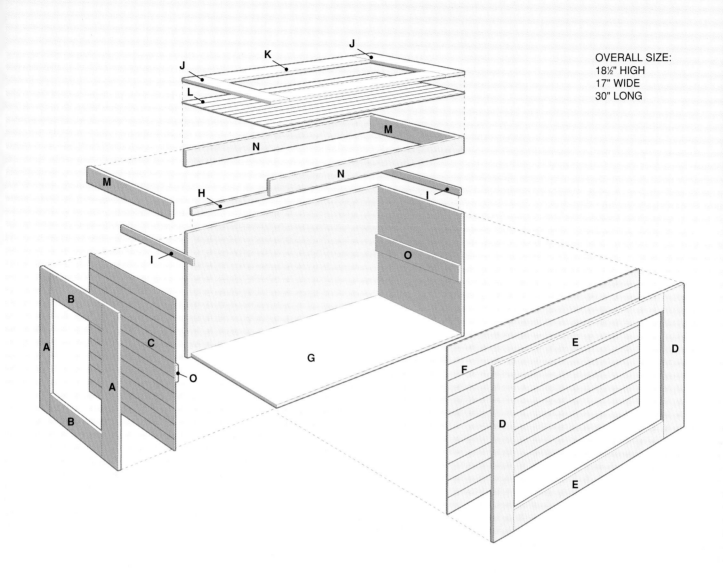

OVERALL SIZE:
18½" HIGH
17" WIDE
30" LONG

Cutting List						Cutting List				
Key	Part	Dimension	Pcs.	Material		Key	Part	Dimension	Pcs.	Material
A	End stile	¾ × 2½ × 16"	4	Pine		I	End lip	½ × 1¼ × 15"	2	Pine molding
B	End rail	¾ × 2½ × 10½"	4	Pine		J	Cover stile	¾ × 2½ × 15½"	2	Pine
C	End panel	¼ × 14½ × 15½"	2	Beaded pine		K	Cover rail	¾ × 2½ × 23½"	2	Pine
D	Side stile	¾ × 2½ × 16"	4	Pine		L	Cover panel	¼ × 15½ × 28½"	1	Beaded pine
E	Side rail	¾ × 2½ × 25"	4	Pine		M	End frame	¾ × 2½ × 15½"	2	Pine
F	Side panel	¼ × 14½ × 28"	2	Beaded pine		N	Side frame	¾ × 2½ × 30"	2	Pine
G	Bottom panel	¾ × 15½ × 28"	1	Plywood		O	Handle cleat	¾ × 2½ × 15"	2	Pine
H	Side lip	½ × 1¼ × 28½"	1	Pine molding						

Materials: Wood glue, #6 × ¾", 1¼" and 1½" wood screws, button plugs, hinges, lid support, latch and finishing supplies.

Note: Measurements reflect the actual size of dimensional lumber.

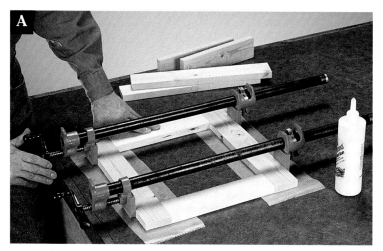

A

Apply glue to the end rails and position them between the end stiles, then clamp in place until dry.

B

Fasten the end panels to the end rails and stiles.

Directions:
Treasure Chest

BUILD THE END & SIDE PANELS. The end and side panels are simple frame constructions with beaded pine paneling fastened to one side. Start the construction by cutting the end stiles (A) and end rails (B) to length from 1 × 3 pine, then sand the edges and surfaces with medium-grit sandpaper. Lay the rails and stiles face-down on a flat surface with the rails between the stiles. Position the rails so their outside edges are flush with the ends of the stiles. Apply wood glue to the ends of the rails and clamp the components **(photo A).** Set the clamped assemblies aside to dry. Next, Cut the end panels (C) to length from beaded pine paneling and sand with medium-grit sandpaper. Unclamp the end rail and stile assemblies and place the end panels on the rail and stile assemblies, flush with the stile edges and ¾" in from the rail edges. Fasten the end panels to the rail and stile assemblies with glue and #6 × ¾" wood screws **(photo B).** Next, cut the side stiles (D) and side rails (E) to length from 1 × 3 pine and sand with medium-grit sandpaper. Glue and clamp the rails and stiles as done previously for the end assemblies and set aside to dry. Next, cut the side panels (F) to size from beaded pine paneling and sand with medium-grit sandpaper. Unclamp the side rail and stile assemblies and place the side panels on the rail and stile assemblies, 1" in from the stile edges and ¾" in from the rail edges. Fasten the side panels to the rails and stile assemblies with glue and #6 × ¾" wood screws.

C

Clamp the end panel assemblies between the side panel assemblies and drill counterbored pilot holes.

ASSEMBLE THE PANELS. Stand an end panel assembly and a side panel assembly on their bottom edges on a flat worksurface. Position the end panel assemblies flush with the ends of the side panel assemblies. Clamp the panels and drill counterbored pilot holes through the side stiles into the end stiles **(photo C).** Be sure to counterbore for wood plugs. Unclamp the panels and join them with glue and #6 × 1½" wood screws. Next, cut the bottom panel (G) to size from ¾"-thick plywood and sand with medium-grit sandpaper.

Attach the end and side frames to the cover panel assembly with glue and counterbored screws.

Fasten the handle cleats to the insides of the end panels with glue and screws.

Position the bottom panel in place and secure it with glue and counterbored wood screws. Cut the side lip (H) and end lip (I) to length from ½ × 1¼" pine stop molding. Be sure to leave one lip slightly shorter to allow room for the lid support. Sand the edges and surfaces with medium-grit sandpaper. Fasten the lips to the end and side rails with their bottom edges ¾" below the top edge of the rails, using glue and #6 × 1¼" wood screws.

BUILD THE COVER. The cover is constructed similarly to the end and side panels and has a frame surrounding it. Cut the cover stiles (J) and cover rails (K) to length from 1 × 3 pine. Sand the edges and surfaces with medium-grit sandpaper. Position, glue and clamp the rails and stiles as done previously for the end and side assemblies. Cut the cover panel (L) to size from beaded pine paneling. Unclamp the cover rail and stile assemblies and lay them facedown on a flat surface. Place the cover panel on the rail and stile assemblies, flush with the stile and rail

edges. Fasten the cover panel to the rail and stile assemblies with glue and #6 × ¾" wood screws. Cut the end frames (M) and side frames (N) to length from 1 × 3 pine to fit the perimeter of the cover panel assembly. Attach the end and side frames to the cover panel assembly with glue and counterbored wood screws **(photo D).** If you choose to have handles on the chest, you'll need handle cleats to mount the handles. Cut the handle cleats (O) to size from 1 × 3 pine and fasten them 7" down from the top of the end lips with glue and 1½" screws driven into the end stiles **(photo E).**

APPLY FINISHING TOUCHES. The easiest way to complete the treasure chest is to apply the finish, then install the hardware. Start by finish-sanding

the entire chest and filling all open counterbore holes with wood plugs. Sand the plugs flush, then apply several coats of quality primer and enamel paint, letting each coat dry thoroughly between applications. When the paint has dried, place the cover on the chest. Mark the hinge locations, on the back side of the chest and cover, 3" in from each end. Attach the hinges to the side frame and side rail with the hinge screw included with the hardware **(photo F).** Install the handles, latch and lid support.

Mount the hinges to the back side of the chest and cover.

Artist Center

*This artist center is a desk, painting easel, seat and storage
center all wrapped into one easy-to-build project.*

CONSTRUCTION MATERIALS

Quantity	Lumber
1	¾" × 4 × 8' plywood
1	¾" × 4 × 4' melamine
1	½" × 4 × 4' plywood

A self-contained artist center is a blessing for parent and child alike. This efficient artist center is a two-part project with a stationary work unit and rolling easel unit. The work unit features a convenient desk and a shelf. In addition to providing a flat surface for painting or drawing, the easel unit has a pull-out tray and a large bin for storing brushes, markers and paper. There's a bench on top of the bin, so children can sit and work at the desk before turning their attention to the painting easel. When not in use, the easel/seat unit fits snugly into the work unit, as a car fits into a garage, taking up a bare minimum of floor space.

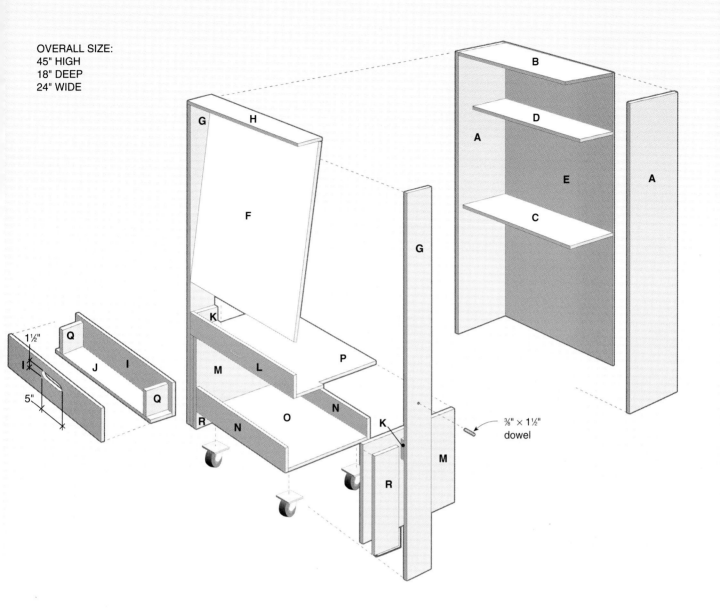

OVERALL SIZE:
45" HIGH
18" DEEP
24" WIDE

³⁄₈" × 1½"
dowel

Cutting List				
Key	Part	Dimension	Pcs.	Material
A	Shelf side	¾ × 13 × 45"	2	Plywood
B	Shelf top	¾ × 13 × 22½"	1	Plywood
C	Desk top	¾ × 12 × 22½"	1	Melamine
D	Top shelf	¾ × 8 × 22½"	1	Melamine
E	Back	½ × 24 × 45"	1	Plywood
F	Easel	¾ × 22½ × 25"	1	Melamine
G	Leg	¾ × 5 × 45"	2	Plywood
H	Hood top	¾ × 5 × 22½"	1	Plywood
I	Tray side	¾ × 4½ × 22⅜"	2	Plywood

Cutting List				
Key	Part	Dimension	Pcs.	Material
J	Tray bottom	¾ × 3½ × 22⅜"	1	Plywood
K	Tray glide	¾ × 2¾ × 4¼"	2	Plywood
L	Seat lip	¾ × 2½ × 22½"	1	Plywood
M	Seat side	¾ × 12¼ × 14¾"	2	Plywood
N	Bin side	¾ × 3½ × 20"	2	Plywood
O	Bin bottom	¾ × 12½ × 20"	1	Plywood
P	Seat	¾ × 15 × 22½"	1	Plywood
Q	Tray cap	¾ × 3½ × 2½"	2	Plywood
R	Spacer	½ × 5 × 12¼"	2	Plywood

Materials: Glue, wood screws (1¼", 2"), 1" wire nails, 3"-dia. casters (4), ⅜"-dia. dowels, pull handle, adjustable glide feet, finishing materials.

Note: Measurements reflect the actual size of dimensional lumber.

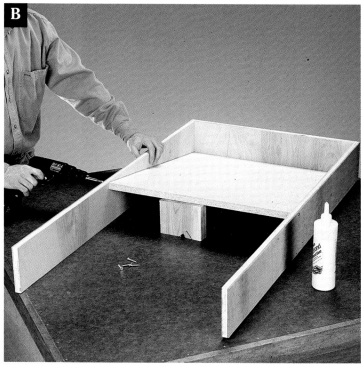

Use a circular saw to cut the melamine for the desk top and top shelf.

Attach the melamine easel between the frame legs so it slants from the seat side of the frame to the easel side.

Directions: Artist Center

BUILD THE WORK UNIT. The stationary work unit is a simple shelf assembly—the shelves are fastened between two tall boards with glue and wood screws to make a simple desk and work station for a child. Cut the shelf sides (A) and shelf top (B) to size from ¾"-thick plywood. Sand all parts with medium sandpaper to smooth out any rough spots. Use a circular saw to cut the desk top (C) and top shelf (D) to size from ¾"-thick melamine-coated particleboard **(photo A).** Mark lines across the shelf sides, 20" and 36" up from the bottom edges. These are the positions of the tops of the desk top and top shelf. Position

the desk top, top shelf and shelf top between the shelf sides. The shelf top should be flush with the top edges of the shelf sides. All three boards should be positioned flush with the back edges of the shelf sides. Use glue and #6 × 2" wood screws to attach the parts—be sure to drill pilot holes for the screws, counterbored so the screw heads can be covered with wood putty. Drive the wood screws through the shelf sides into the ends of the shelf top, desk top and top shelf. Cut the back (E) to size, and fasten it to the back of the work center with 1" wire nails. As you attach the back, measure diagonally from corner to corner to make sure the work unit is in square. If the measurements differ, the unit is out of square: apply pressure to one side to push it back into square.

BUILD THE EASEL. The easel is a piece of melamine board with a plywood frame. Cut the easel (F), legs (G) and hood top (H) to size. Fasten the hood top between the tops of the legs with glue and counterbored wood screws, driven through the legs and into the ends of the hood top. Lay the legs on a flat worksurface, and position the easel between the legs. The easel should slant from the back of the frame toward the front. Position it so the front edge of the easel is flush with the front edge of the hood top. Drive a #6 × 2" wood screw through the top of each leg and into the edges of the easel—make sure to drill counterbored pilot holes first. Drive only one screw at each side, then flip the frame so the back edges are pointing up. Slip a wood spacer under the easel,

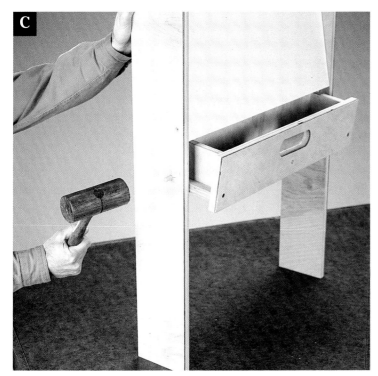

Insert dowels into guide holes in the frame legs to prevent the tray from pulling all the way out of the frame.

INSTALL THE TRAY. The easel tray fits into the frame, just below the easel, for convenient storage of markers and paints. It slides on two plywood tray glides that are attached to the insides of the legs. Dowels are inserted through guide holes in the legs after the tray is installed, creating stops so the tray cannot be pulled out completely. Draw reference lines across the inside faces of both frame legs, 15¾" up from the bottoms. Cut the plywood tray glides (K) to size, then attach them to the insides of the frame legs, just below the reference lines. Use counterbored #6 × 1¼" wood screws driven through the glides and into the legs. Mark points on the outside face of each leg, 16¾" up from the bottom and 1⅛" in from the front edge of each leg. Drill ⅜"-dia.

TIP

Wood dowels are often slightly thinner than their stated diameter. Drill a test hole in a piece of scrap with a drill bit that is the same size as the dowels are supposed to be. Insert the dowel to make sure it fits snugly. If the dowel is loose, switch to a slightly smaller bit (not more than ¹⁄₁₆" smaller).

propping it up until the back surface is flush with the back edges of the legs. Drill counterbored pilot holes, then drive screws through the legs and into the bottoms of the easel **(photo B).** Add at least one more screw to each side, midway up the edge of the easel.

BUILD THE EASEL TRAY. Cut the tray sides (I), tray bottom (J), tray glides (K) and tray caps (Q) to size. Cut a centered, 1½"-wide × 5"-long notch in one tray side (see *Diagram,* page 93) to use as a handle grip for pulling the tray out. Start the notch 1" from the top edge of the tray side—cut the ends with a 1½"-dia. hole saw mounted on your electric drill, then join the holes with a jig saw to complete the handle grip. Mark a line across the notched tray side, 1½" up from the bottom edge. Use glue and counterbored wood screws to attach the tray bottom with its

top face at the line. Use glue and wood screws to fasten the other tray side to the tray bottom. Glue the tray caps into position between the tray sides, ¾" in from the ends of the tray. Secure the tray caps in place with tape until the glue has set.

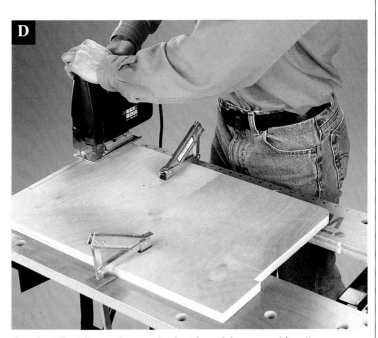

Cut the ½"-wide notches on both sides of the seat with a jig saw.

guide holes for the dowel stops at these points (see *Tip*, previous page). Slide the tray between the legs so it rests on the tray glides. Cut and insert a ⅜ × 1½"-long dowel into the guide holes in the legs **(photo C).** Do not glue the dowels if you wish to have the option of removing the tray for cleaning.

BUILD THE BIN. The bin is a storage box that fits in the easel frame between the bottoms of the frame legs. Cut the seat sides (M), bin sides (N) and bin bottom (O). Fasten the bin bottom between the bin sides with glue and counterbored wood screws, making sure their bottom edges are flush. Attach the bin bottom and sides between the seat sides, 3" up from the bottom of the seat sides. Cut the seat (P) and seat lip (L). Draw a ½"-wide × 10¾"-long notch at the sides of the seat (see *Diagram,* page 93), starting at the front end. The notches reduce the width of the seat so it can fit easily into the work unit when not in use. Cut out the notches with a jig saw **(photo D),** then sand the edges smooth. Use a compass

Attach plywood spacers to fill the gap between the seat and the easel frame before the parts are screwed together.

to draw 1"-radius roundovers at each front corner of the seat. Cut the curves with a jig saw, and sand smooth. Fasten the seat lip to the back edge of the seat, making sure the bottoms of the pieces are flush. Use #6 × 2" wood screws, driven through the seat lip and into the back edge of the seat. Fasten the seat to the tops of the seat sides. Make sure the back of the seat lip is flush with the back edges of the seat sides.

ATTACH THE SEAT & BIN. The seat and bin assembly is fastened between the frame legs. Cut the spacers (R) to size from ½"-thick plywood, and attach them to the outside faces of the seat sides, with the front and bottom edges flush **(photo E).** Position the seat and bin assembly between the legs so that the seat side bottoms are flush with the leg bottoms. Attach the assemblies with glue and #6 × 2" wood screws, driven through the legs and into the spacers.

APPLY FINISHING TOUCHES. Fill all counterbores and exposed plywood edges with wood putty, then sand all the wood surfaces smooth. Remove the tray, then prime and paint all wood surfaces (we used a hard enamel paint because it is washable). Do not paint the melamine surfaces. After the paint has dried, attach 3"-dia. casters to the undersides of the bin bottom, as far toward the front and back edges as possible **(photo F).** Attach adjustable glide feet to the bottom of the work unit so it can be raised or lowered to set clearance for the seat when the the artist center is not in use. Reinsert the tray and drive the dowel stops back into the guide holes to keep the tray from falling out. Finally, attach a drawer pull to the outside face of the seat lip. This handle makes it easy to pull the easel unit away from the work center.

Attach 3"-dia. casters to the bottom of the easel-back seat.